Emerging Micro Manufacturing Technologies and Applications

Emerging Micro Manufacturing Technologies and Applications

Editor

Nikolaos Tapoglou

MDPI • Basel • Beijing • Wuhan • Barcelona • Belgrade • Manchester • Tokyo • Cluj • Tianjin

Editor
Nikolaos Tapoglou
Department of Industrial
Engineering and
Management
International Hellenic
University
Thessaloniki
Greece

Editorial Office
MDPI
St. Alban-Anlage 66
4052 Basel, Switzerland

This is a reprint of articles from the Special Issue published online in the open access journal *Micromachines* (ISSN 2072-666X) (available at: www.mdpi.com/journal/micromachines/special_issues/micro_manufacturing_technologies).

For citation purposes, cite each article independently as indicated on the article page online and as indicated below:

LastName, A.A.; LastName, B.B.; LastName, C.C. Article Title. *Journal Name* **Year**, *Volume Number*, Page Range.

ISBN 978-3-0365-8213-9 (Hbk)
ISBN 978-3-0365-8212-2 (PDF)

Contents

About the Editor . . . vii

Nikolaos Tapoglou
Editorial for the Special Issue on Emerging Micro Manufacturing Technologies and Applications
Reprinted from: *Micromachines* **2023**, *14*, 1248, doi:10.3390/mi14061248 . . . 1

Petr Hauschwitz, Jan Brajer, Danijela Rostohar, Jaromír Kopeček, Tomáš Mocek and Martin Cimrman et al.
Anti-Reflection Nanostructures on Tempered Glass by Dynamic Beam Shaping
Reprinted from: *Micromachines* **2021**, *12*, 289, doi:10.3390/mi12030289 . . . 5

Marta Maria Kluba, Jian Li, Katja Parkkinen, Marcus Louwerse, Jaap Snijder and Ronald Dekker
Cavity-BOX SOI: Advanced Silicon Substrate with Pre-Patterned BOX for Monolithic MEMS Fabrication
Reprinted from: *Micromachines* **2021**, *12*, 414, doi:10.3390/mi12040414 . . . 11

Kai Li, Zhenyu Zhao, Houming Zhou, Hao Zhou, Jie Yin and Wei Zhang et al.
Numerical Simulation of Effect of Different Initial Morphologies on Melt Hydrodynamics in Laser Polishing of Ti6Al4V
Reprinted from: *Micromachines* **2021**, *12*, 581, doi:10.3390/mi12050581 . . . 23

Jun Cao, Junchao Zhang, Baoan Wu, Huiyi Tang, Changchun Lv and Kexing Song et al.
Study on Manufacturing Technology of Ag 8.5Au 3.5Pd Fine Alloy Wire
Reprinted from: *Micromachines* **2021**, *12*, 938, doi:10.3390/mi12080938 . . . 43

Zengxing Zhang, Guohua Liu and Kaiying Wang
Fabrication of Needle-like Silicon Nanowires by Using a Nanoparticles-Assisted Bosch Process for Both High Hydrophobicity and Anti-Reflection
Reprinted from: *Micromachines* **2021**, *12*, 1009, doi:10.3390/mi12091009 . . . 59

Ki Seok Kim, You-Jin Ji, Ki-Hyun Kim, Ji-Eun Kang, Albert Rogers Ellingboe and Geun Young Yeom
Deposition of Very-Low-Hydrogen-Containing Silicon at a Low Temperature Using Very-High-Frequency (162 MHz) SiH_4 Plasma
Reprinted from: *Micromachines* **2022**, *13*, 173, doi:10.3390/mi13020173 . . . 69

Seitaro Kumamoto, Souichiro Fukuyama, Seiya Nagano, Keiichiro Yasuda, Yusuke Kitamura and Masaaki Iwatsuki et al.
Fabrication of Three-Dimensionally Deformable Metal Structures Using Precision Electroforming
Reprinted from: *Micromachines* **2022**, *13*, 1046, doi:10.3390/mi13071046 . . . 77

JagadeeswaraRao Maddu, Buschaiah Karrolla, Riyaaz Uddien Shaik, Hassan Elahi and Krishnaiah Arkanti
Evaluation of Bronze Electrode in Electrical Discharge Coating Process for Copper Coating
Reprinted from: *Micromachines* **2023**, *14*, 136, doi:10.3390/mi14010136 . . . 93

Andrey Yankin, Gaini Serik, Saniya Danenova, Yerassyl Alipov, Ali Temirgali and Didier Talamona et al.
Optimization of Fatigue Performance of FDM ABS and Nylon Printed Parts
Reprinted from: *Micromachines* **2023**, *14*, 304, doi:10.3390/mi14020304 . . . 111

Hongliang Zhou, Yingchong Zhang, Jun Cao, Chenghao Su, Chong Li and Andong Chang et al.
Research Progress on Bonding Wire for Microelectronic Packaging
Reprinted from: *Micromachines* **2023**, *14*, 432, doi:10.3390/mi14020432 . . . . **127**

About the Editor

Nikolaos Tapoglou

Dr Nikolaos Tapoglou is an Assistant Professor in the Department of Industrial Engineering and Management of the International Hellenic University Thessaloniki, Greece (IHU). He received his Diploma, M.Sc., and PhD with a specialisation in manufacturing processes from the department of Production Engineering and Management of the Technical University of Crete.

Prior to joining IHU, he led the emerging machining technology team in the Advanced Manufacturing Research Centre (AMRC) of the University of Sheffield as a Technical Fellow.

His research focuses on additive manufacturing processes, emerging machining technologies, cloud manufacturing, the study and optimisation of advanced manufacturing processes, manufacturing simulation focusing on the calculation of in-process characteristics and manufacturing informatics. He has co-authored over 30 journal and conference publications and has worked on more than 20 research projects.

micromachines

Editorial

Editorial for the Special Issue on Emerging Micro Manufacturing Technologies and Applications

Nikolaos Tapoglou

Industrial Engineering and Management Department, International Hellenic University, 57001 Thessaloniki, Greece; ntapoglou@iem.ihu.gr

In recent years, the field of micromachining has gained a lot of traction owing to the drive towards lightweighting, electrification, and sustainability. Industrial sectors that have shown an increasing interest in micromachining include the medical, space, aerospace, and consumer electronics fields. Research in academia has focused on the experimental investigation of micromachining and additive manufacturing processes and in particular, on laser-based manufacturing technologies. In addition, numerical and finite element models have been developed to predict the performance of micromachined parts. Over the last few years, a series of manufacturing processes have emerged in the macro manufacturing sector that have shown great potential in the improvement of these processes; however, their use on the micro scale has not been thoroughly modeled and understood. Moreover, a series of processes developed to address challenges in micro-manufacturing have been emerging as a parallel thread. Accordingly, this Special Issue showcases nine original research papers and one state-of-the-art review in the fields of manufacturing micro-electronic devices, surface property modification, and additive manufacturing.

In particular, Hauschwitz et al. [1] presented a novel technique for the modification of the reflectivity and surface topography of tempered glass using an ultrashort, pulsed laser. By utilizing a dynamic beam shaping and a galvanometer scanning head, the laser beam was divided into a matrix of beamlets, allowing for the fast and flexible fabrication of a sub-wavelength ripple structure on the surface. The study showed that reflected intensity reduced by up to 75% while maintaining 90% of transparency.

Kluba et al. [2] presented a novel type of substrate for a Silicon on Insulator wafer, which contained a patterned, buried oxide layer that can simplify the fabrication of MEMS devices with complex geometry and added functionality. The authors successfully demonstrated the application of the cavity-BOX SOI substrate in the fabrication of a deep brain stimulation (DBS) demonstrator with a length of 18 mm and a diameter of 1.39 mm.

Li et al. [3] developed a 2D model of the molten flow behavior during laser polishing, demonstrating the complex evolution of melt hydrodynamics involving heat conduction, thermal convection, thermal radiation, melting, and solidification. The model developed was able to predict that the morphological evolution of different surfaces from rough to smooth in laser polishing could guide the optimization of polishing parameters such as laser power and scanning speed. The experimental results showed a good correlation between the experimental and simulated results with an error ranging from 8.3 to 14.3%.

Cao et al. [4] investigated the performance of a Ag-8.5Au-3.5Pd alloy wire after cold deformation and annealing, using SEM, strength, and resistivity testing. The experimental campaign showed that the strength of the wire increases with the increase in the deformation rate, and the resistivity decreases with the increase in the annealing temperature. The mechanical performance of the alloy wire was improved at an annealing temperature of 500 °C. The surface quality is high when the tension range is 2.5–3.0 g. However, when the annealing temperature increased to 550 °C, the grain size growth led to a decrease in the mechanical performance.

Citation: Tapoglou, N. Editorial for the Special Issue on Emerging Micro Manufacturing Technologies and Applications. *Micromachines* **2023**, *14*, 1248. https://doi.org/10.3390/mi14061248

Received: 7 June 2023
Accepted: 10 June 2023
Published: 14 June 2023

Zhang et al. [5] presented a modified Bosch etching process to create silicon nanowires with a controllable density and high aspect ratios. The use of Au nanoparticles as a hard mask resulted in an anti-reflection property with a reflectance value below 2% in a broad light wave range and a near-unity reflectance below 3% in the range from 220 to 2600 nm. Additionally, the nanowire array demonstrated super-hydrophobic behavior without any hydrophobic chemical treatment.

Kim et al. [6] investigated the deposition of low-hydrogen-containing amorphous silicon (a-Si) using a plasma system with multi-split electrodes. Based on their experimental research, they concluded that increasing the RF power of the plasma led to decreased hydrogen content in the deposited film and a decrease in impurities such as carbon and oxygen. When crystalized under a UV lamp, a-Si exhibited improved crystallinity, confirmed by Raman spectroscopy and HR-TEM.

Kumamoto et al. [7] introduced a technique for fabricating metal structures that could be deformed from two-dimensional to three-dimensional shapes in response to hydrodynamic forces using photolithography and electroforming. The resulting structures had an average film thickness of 12.9 μm, a hardness of 600 HV, and a slit width of 7.9 μm.

Maddu et al. [8] evaluated the use of a bronze electrode for depositing copper material on titanium alloy using electrical discharge machining (EDM). The input parameters of current, Ton, Toff, and preheating substrate temperature were optimized using the Taguchi experimental design and the TOPSIS technique. The study found that input parameters of current 8 Amp, Ton 440 μs, Toff 200 μs, temperature 300 °C, and a quenching medium of castor oil provided the optimum response of an MDR of 0.00506 g/min, EWR of 0.00462 g/min, CT of 40.2 μm, and SCD of 19.4×10^7 μm^2.

Yakin et al. [9] aimed to optimize Fused Deposition Modeling (FDM) printing parameters for ABS and Nylon to improve fatigue resistance. The methodology involved experimental study and finite element analysis. The results showed that Nylon performed better than ABS, and the 'tri-hexagon' structure with a nozzle diameter of 0.2 mm resulted in the highest fatigue life for both materials.

Finally, Zhou et al. [10] discussed the history and development of bonding wires in microelectronic packaging, specifically comparing the properties and applications of Au, Cu, and Ag bonding wires. The paper highlighted the benefits and challenges of each material and suggested future research focuses on understanding the bonding mechanism, developing environmentally friendly surface treatment technology, exploring multi-component microalloy doping processes, and optimizing wire loop profiles for improved reliability.

Acknowledgments: The guest editor would like to thank the authors for submitting their excellent contributions to this Special Issue. Furthermore, the present Special Issue would not have been possible without the expert reviewers who carefully evaluated the manuscripts and provided helpful comments and suggestions for improvements. A special thank you goes to the editors and the MDPI team for their outstanding management of this Special Issue.

Conflicts of Interest: The author declares no conflict of interest.

References

1. Hauschwitz, P.; Brajer, J.; Rostohar, D.; Kopeček, J.; Mocek, T.; Cimrman, M.; Chyla, M.; Smrž, M. Anti-Reflection Nanostructures on Tempered Glass by Dynamic Beam Shaping. *Micromachines* **2021**, *12*, 289. [CrossRef] [PubMed]
2. Kluba, M.M.; Li, J.; Parkkinen, K.; Louwerse, M.; Snijder, J.; Dekker, R. Cavity-BOX SOI: Advanced Silicon Substrate with Pre-Patterned BOX for Monolithic MEMS Fabrication. *Micromachines* **2021**, *12*, 414. [CrossRef] [PubMed]
3. Li, K.; Zhao, Z.; Zhou, H.; Zhou, H.; Yin, J.; Zhang, W.; Zhou, G. Numerical Simulation of Effect of Different Initial Morphologies on Melt Hydrodynamics in Laser Polishing of Ti6Al4V. *Micromachines* **2021**, *12*, 581. [CrossRef] [PubMed]
4. Cao, J.; Zhang, J.; Wu, B.; Tang, H.; Lv, C.; Song, K.; Yang, G.; Cui, C.; Gao, Y. Study on Manufacturing Technology of Ag-8.5Au-3.5Pd Fine Alloy Wire. *Micromachines* **2021**, *12*, 938. [CrossRef] [PubMed]
5. Zhang, Z.; Liu, G.; Wang, K. Fabrication of Needle-Like Silicon Nanowires by Using a Nanoparticles-Assisted Bosch Process for Both High Hydrophobicity and Anti-Reflection. *Micromachines* **2021**, *12*, 1009. [CrossRef] [PubMed]
6. Kim, K.S.; Ji, Y.-J.; Kim, K.-H.; Kang, J.-E.; Ellingboe, A.R.; Yeom, G.Y. Deposition of Very-Low-Hydrogen-Containing Silicon at a Low Temperature Using Very-High-Frequency (162 MHz) SiH4 Plasma. *Micromachines* **2022**, *13*, 173. [CrossRef] [PubMed]

7. Kumamoto, S.; Fukuyama, S.; Nagano, S.; Yasuda, K.; Kitamura, Y.; Iwatsuki, M.; Baba, H.; Ihara, T.; Nakanishi, Y.; Nakashima, Y. Fabrication of Three-Dimensionally Deformable Metal Structures Using Precision Electroforming. *Micromachines* **2022**, *13*, 1046. [CrossRef] [PubMed]
8. Maddu, J.; Karrolla, B.; Shaik, R.U.; Elahi, H.; Arkanti, K. Evaluation of Bronze Electrode in Electrical Discharge Coating Process for Copper Coating. *Micromachines* **2023**, *14*, 136. [CrossRef] [PubMed]
9. Yankin, A.; Serik, G.; Danenova, S.; Alipov, Y.; Temirgali, A.; Talamona, D.; Perveen, A. Optimization of Fatigue Performance of FDM ABS and Nylon Printed Parts. *Micromachines* **2023**, *14*, 304. [CrossRef] [PubMed]
10. Zhou, H.; Zhang, Y.; Cao, J.; Su, C.; Li, C.; Chang, A.; An, B. Research Progress on Bonding Wire for Microelectronic Packaging. *Micromachines* **2023**, *14*, 432. [CrossRef] [PubMed]

micromachines

Communication

Anti-Reflection Nanostructures on Tempered Glass by Dynamic Beam Shaping

Petr Hauschwitz [1,2,*], Jan Brajer [1], Danijela Rostohar [1], Jaromír Kopeček [3], Tomáš Mocek [1], Martin Cimrman [1,2], Michal Chyla [1] and Martin Smrž [1]

1 HiLASE Centre, Institute of Physics, Czech Academy of Sciences, Za Radnici 828, 25241 Dolni Brezany, Czech Republic; Jan.Brajer@fzu.cz (J.B.); Danijela.Rostohar@fzu.cz (D.R.); Tomas.Mocek@fzu.cz (T.M.); Martin.Cimrman@fzu.cz (M.C.); Michal.Chyla@fzu.cz (M.C.); Martin.Smrz@fzu.cz (M.S.)

2 Faculty of Nuclear Sciences and Physical Engineering, Czech Technical University in Prague, Brehova 7, 11519 Prague, Czech Republic

3 Institute of Physics of the Czech Academy of Sciences, Na Slovance 2, 18221 Prague, Czech Republic; Jaromir.Kopecek@fzu.cz

* Correspondence: petr.hauschwitz@hilase.cz

Abstract: Reflectivity and surface topography of tempered glass were modified without any thermal damage to the surroundings by utilizing 1.7 ps ultrashort pulsed laser on its fundamental wavelength of 1030 nm. To speed up the fabrication, a dynamic beam shaping unit combined with a galvanometer scanning head was applied to divide the initial laser beam into a matrix of beamlets with adjustable beamlets number and separation distance. By tuning the laser and processing parameters, reflected intensity can be reduced up to 75% while maintaining 90% of transparency thus showing great potential for display functionalization of mobile phones or laptops.

Keywords: multi-beam micromachning; functionalization; tempered glass; gloss reduction; anti-reflection

Citation: Hauschwitz, P.; Brajer, J.; Rostohar, D.; Kopeček, J.; Mocek, T.; Cimrman, M.; Chyla, M.; Smrž, M. Anti-Reflection Nanostructures on Tempered Glass by Dynamic Beam Shaping. *Micromachines* **2021**, *12*, 289. https://doi.org/10.3390/mi12030289

Academic Editor: Nikolaos Tapoglou

Received: 12 February 2021
Accepted: 8 March 2021
Published: 9 March 2021

Publisher's Note: MDPI stays neutral with regard to jurisdictional claims in published maps and institutional affiliations.

1. Introduction

The value of common everyday products like mobile phones or laptops can be compromised with a poor anti-reflection property of a covering display glass making it unreadable under direct sun. Many methods have been used for the fabrication of antireflection or gloss reduction surface structures including lithography [1], sol-gel [2], etching [3] and multi-stage deposition methods [4]. However, these methods are usually complicated with several steps, long processing times or require chemicals and thus are not environmentally friendly. On the other hand, laser surface micro/nanostructuring offers a flexible, fast and environmentally friendly approach for precise and efficient fabrication of desired micro/nanogeometry in a single step [5].

Laser-made antireflection surface structures are often composed of periodic sub-wavelength structures [6], laser-induced periodic surface structures (LIPSS) [7] or blind microholes [8] which can alter the refractive index to capture the incoming light. High-resolution of required structures results in longer fabrication times, as well as delicate power-handling close to damage threshold to ensure high-quality processing without thermal damage to the surroundings. To improve the fabrication time and use the laser source more efficiently, multi-beam processing can be applied [9]. However, commonly used diffractive beamsplitters may not provide sufficient freedom in adjusting the spacing between beamlets during process optimization.

In this paper, an ultrashort pulsed laser is used to modify the surface reflectivity by reliable production of sub-wavelength surface structures on tempered glass sample, commonly used as a protective glass layer on displays of mobile phones, smartwatches or laptops. The fabrication speed and process efficiency are improved by a dynamic beam shaping unit combined with a galvanometer scanner allowing to freely divide the beam and adjust a diffraction pattern in real-time. Thus, demonstrating the production of anti-reflection surfaces by dynamic multi-beam glass processing for the first time.

2. Materials and Methods

Tempered glass plates with dimensions of 120 × 60 mm and thickness of 500 μm were cleaned in an ultrasonic bath with ethanol before the laser treatment with ultrashort pulsed laser system Perla B (HiLASE Center) emitting 1.7 ps pulses at 1030 nm with the repetition rate of 1 kHz. The output beam was guided into the dynamic beam shaping unit FBS G3 (Pulsar Photonics GmbH) equipped with a spatial light modulator (SLM, Hamamatsu Photonics) and galvanometer scanner (intelliSCAN III 14, Scanlab) with the maximum marking speed of 2 m/s. The beam was focused on a sample with 100 mm telecentric F-theta lens resulting in a spot diameter of 20 μm.

The input beam wavefront was modified by SLM resulting in desired diffractive patterns which are generated in real-time by uploading the pre-calculated computer-generated holograms (phase-masks) on SLM. The final patterns on a sample surface were in a form of orthogonal dot matrixes with the highest possible number of beamlets to fit in the LIDT limit of SLM which is 0.5 mJ for picosecond pulses.

The geometry of fabricated structures was investigated with a scanning electron microscope (SEM) Tescan FERA 3 and laser scanning confocal microscope Olympus OLS5000. Specular reflection and transmission were measured using a spectrometer Green Wave VIS and a tungsten krypton lamp, both coupled into an optical fiber. Precision home-made holders were applied to hold these fibers with respect to the required angles of incidence. The specular reflection and transmission were firstly measured on the plane untreated sample for reference. In the second step, the reflection and transmission were measured on laser patterned samples. By comparing these values, drop in reflection and transmission of the patterned surface is measured in %.

3. Results and Discussion

In the first experimental step, the pulse energy and a number of pulses had to be carefully adjusted to selectively modify glass surface in well-defined spot regions without glass cracking and thermal damage. Glass cracking was not observed for pulse energies below 200 μJ (64 J/cm^2) for a single pulse exposition. However, with a slightly increased pulse count, the glass surface starts to crack (not shown). The optimal pulse energy was determined as 40 μJ (13 J/cm^2), slightly above the single-pulse damage threshold of 10 J/cm^2, as the single pulse can still modify the glass surface and no cracks are observed up to 10 consecutive laser pulses.

Following these observations, the computer-generated hologram responsible for a diffraction pattern dividing the initial beam into the matrix of 3 × 3 beamlets was uploaded on the SLM plate (Figure 1a). The pulse energy of the initial beam was then increased up to 420 μJ providing ~40 μJ in each beamlet (85% SLM efficiency). The final pattern of 3 × 3 matrix fabricated on a glass surface in a single pulse is depicted in a microscope image in Figure 1b.

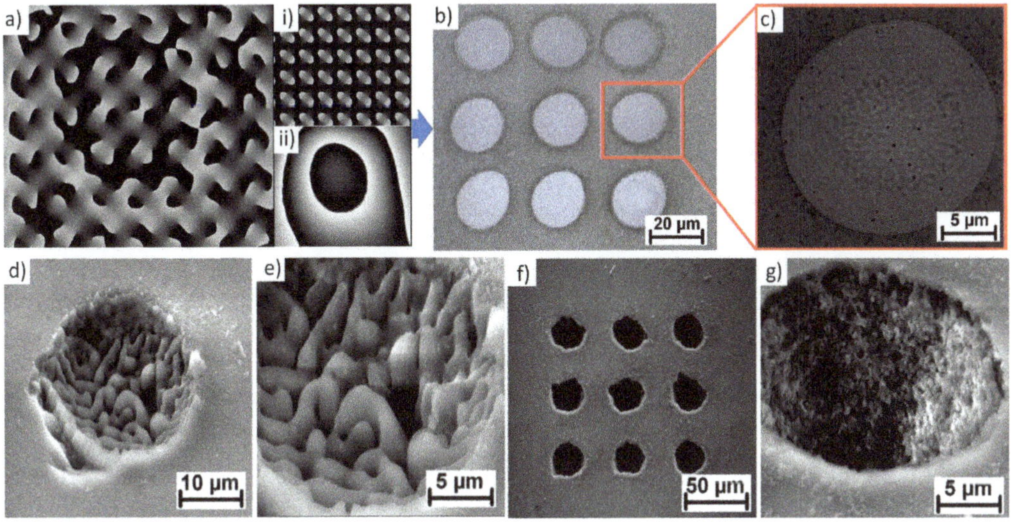

Figure 1. (**a**) Computer generated hologram composed of hologram dividing the initial beam into 3 × 3 matrix (**i**) and correction file compensating SLM flatness (**ii**); (**b**) Final pattern on a glass sample after single pulse exposition with 40 µJ; (**c**) Spot detail indicating sub-wavelength structures in the center part; (**d**) SEM image of a microhole formed after 5 consecutive laser pulses with a pulse energy of 40 µJ, (**e**) detail on a ripple structure inside; (**f**) Microholes drilled with 10 consecutive laser pulses, (**g**) microhole detail with nano-scale features formed on side walls.

As shown in Figure 1c, an indication of a subwavelength structure was observed in a center of the spot. Therefore, a detail SEM analysis was carried out revealing the ripple structure inside each spot (Figure 1d,e). The lateral spacing of ripple structures is ~900 nm similar to low-frequency laser-induced periodic surface structures [10].

For a single pulse exposition, the ripple structure is formed only in the center part of the spot covering ~60% of the spot diameter. With the increasing number of pulses, ripples are extending to the edges of the spot covering the whole spot in the case of 5 consecutive pulses (Figure 1d,e). With the higher number of pulses, holes can be drilled inside the material (Figure 1f) showing a very high edge quality and nano scale features on the side walls (Figure 1g).

In the following experimental step, larger areas of 15 × 15 mm were patterned with the optimal pulse energy level of 40 µJ (13 J/cm^2) and by changing the spot separation distance (HD) in the range of 17 µm to 25 µm and a number of pulses (N) in the range of 1 to 5 pulses. The pitch between beamlets in 3 × 3 matrix was always adjusted to fit with the corresponding HD by uploading a new hologram on SLM.

As can be observed in Figure 2, different anti-reflection properties can be reached by altering the spot distance and the number of pulses.

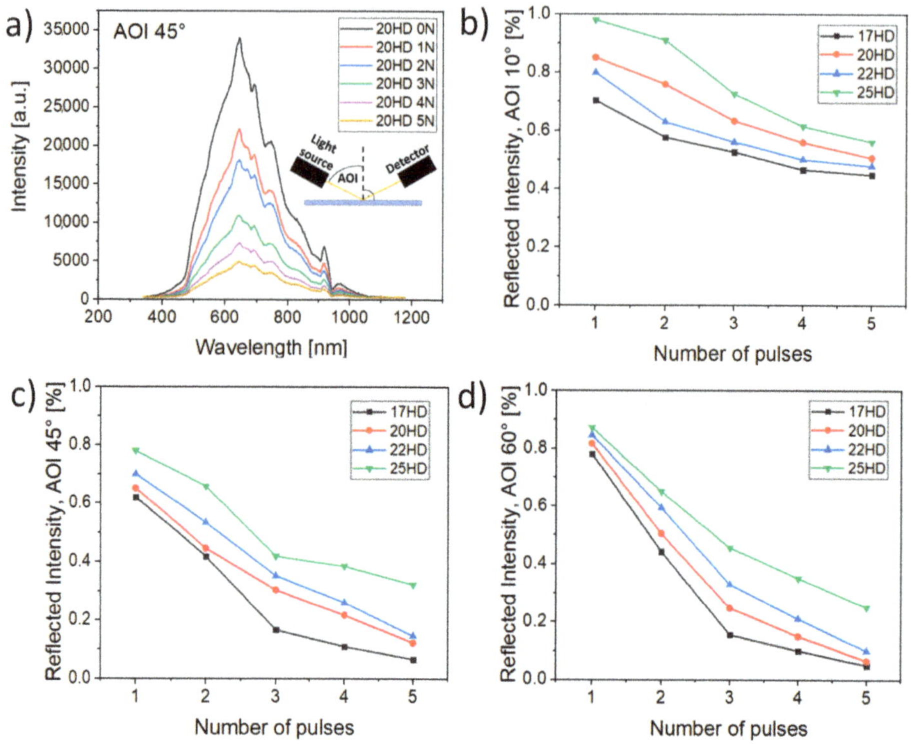

Figure 2. (**a**) Example of reflected intensity curve for an angle of incidence (AOI) = 45° and a different number of applied laser pulses (N); (**b–d**) Sample reflection in dependence on the applied number of pulses (N) and different spot separation distance (HD) for the AIO = 10°, 45°, 60°.

This phenomenon might be related to the complex micro and nano topography of the surface which may result in light trapping or diffraction. As it was shown in earlier studies [11], complex micro and nano topography can enhance the absorption of light by multiple reflections and multiple absorptions compared to that of a perfectly flat surface. Similarly, A.Y. Vorobyev et al. [12] observed the decrease in reflection on nanostructured surface features in comparison to that of a polished surface and Korolkov et al. [13] demonstrated the production of antireflective coatings by laser induced nanostructures on metals.

With the increase in the number of pulses, thus with a higher percentage of ripples covering each spot the higher percentage of the incoming light is diffracted or trapped in between surface features. As a result, the reflected intensity can be decreased by more than 50% in the case of AIO = 10° and more than 90% in the case of AIO = 45° and 60°.

The reflectivity can be also tuned by the spot separation distance. Generally, the closer are the spots, the lower is the reflected intensity as the amount of untreated material decreases. However, the transparency of the structured glass surface decreases as well with the smaller spot distance and the higher number of applied pulses (Figure 3). Therefore, there is a tradeoff between reflectivity and transparency. The detail results of transparency are shown in Figure 3 with a real photograph of structured glass samples.

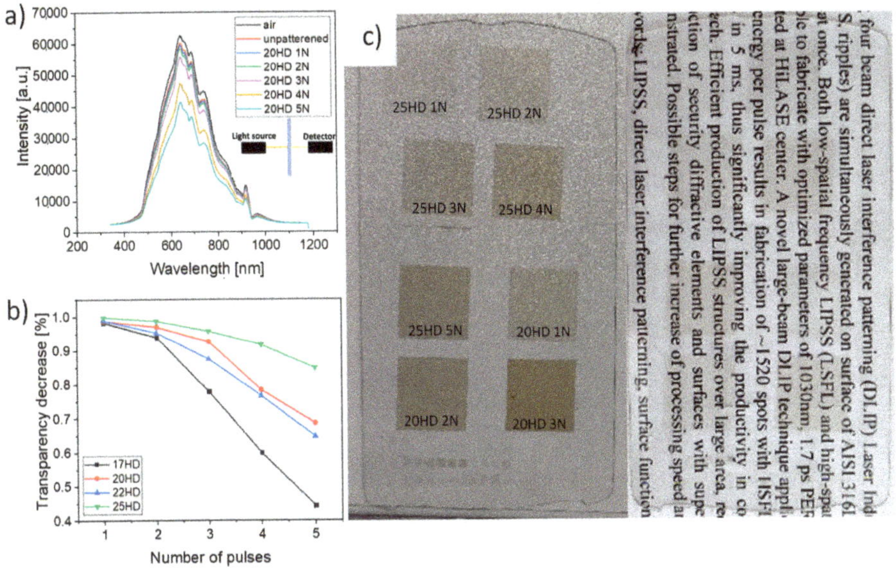

Figure 3. (**a**) Example of transmitted intensity curves for 20HD and different number of applied laser pulses (N); (**b**) Decrease in transparency in dependence on the applied number of pulses (N) and different spot separation distance (HD); (**c**) Photograph of the glass sample on a white table desk and printed paper.

Following these findings in Figures 2 and 3, the best antireflection results with a high level of transparency above 90%, can be achieved with 20 µm spot separation distance and 3 pulses as it provides the highest drop in reflected intensity, down to 63%, 30% and 25% for AOI of 10°, 45° and 60°, respectively. Thus, showing great potential for display cover glass applications.

4. Conclusions

The combination of ultrashort pulse laser and SLM allows fast and flexible fabrication of sub-wavelength ripple structure on a tempered glass surface. The initial laser beam was split into 3 × 3 matrix with adjustable pitch distance increasing the fabrication time nine folds compared to the single beam approach. By adjusting the spot separation distance and number of pulses, different anti-reflection properties can be reached. Up to 75% drop in reflected intensity under AOI = 60° was demonstrated with more than 90% of transparency, thus showing great potential as an alternative to mobile phone or laptop display modification techniques.

Author Contributions: Conceptualization, P.H.; Methodology, P.H.; Validation, P.H., J.B., D.R., J.K., T.M., M.C. (Martin Cimrman), M.C. (Michal Chyla), M.S.; Investigation, P.H., J.K.; Resources, J.B., D.R., J.K., T.M., M.C. (Martin Cimrman), M.C. (Michal Chyla), M.S.; Writing—Original Draft, P.H.; Writing—Review & Editing, P.H.; Visualization, P.H., J.B., D.R., J.K., T.M., M.C. (Martin Cimrman), M.C. (Michal Chyla), M.S.; Project administration, J.B., P.H., T.M.; Funding acquisition, T.M. All authors have read and agreed to the published version of the manuscript.

Funding: This work was co-financed by the European Regional Development Fund and the state budget of the Czech Republic (project HiLASE CoE: Grant No. CZ.02.1.01/0.0/0.0/15_006/0000674) and from the European Union's Horizon 2020 research and innovation programme (Grant agreement NO 739573) and by the Ministry of Education, Youth and Sports of the Czech Republic (Programme NPU I Project No. LO1602).

Conflicts of Interest: The authors declare no conflict of interest. The funders had no role in the design of the study; in the collection, analyses, or interpretation of data; in the writing of the manuscript, or in the decision to publish the results.

References

1. Han, K.-S.; Shin, J.-H.; Yoon, W.-Y.; Lee, H. Enhanced performance of solar cells with anti-reflection layer fabricated by nano-imprint lithography. *Sol. Energy Mater. Sol. Cells* **2011**, *95*, 288–291. [CrossRef]
2. Chen, D. Anti-reflection (AR) coatings made by sol–gel processes: A review. *Sol. Energy Mater. Sol. Cells* **2001**, *68*, 313–336. [CrossRef]
3. Zhang, J.-F.; Zhu, Y.; Li, Z.-H.; Zhou, W.; Zheng, J.-C.; Yang, D.-Q. Preparation of anti-reflection glass surface with self-cleaning and anti-dust by ammonium hydroxide hydrothermal method. *Mater. Express* **2015**, *5*, 280–290. [CrossRef]
4. Wang, W.-C.; Lin, C.-W.; Chen, H.-J.; Chang, C.-W.; Huang, J.-J.; Yang, M.-J.; Tjahjono, B.; Huang, J.-J.; Hsu, W.-C.; Chen, M.-J. Surface Passivation of Efficient Nanotextured Black Silicon Solar Cells Using Thermal Atomic Layer Deposition. *ACS Appl. Mater. Interfaces* **2013**, *5*, 9752–9759. [CrossRef] [PubMed]
5. Hauschwitz, P.; Jochcová, D.; Jagdheesh, R.; Rostohar, D.; Brajer, J.; Kopeček, J.; Cimrman, M.; Smrž, T.; Mocek, A. Lucianetti, Towards rapid large-scale LIPSS fabrication by 4-beam ps DLIP. *Opt. Laser Technol.* **2021**, *133*, 106532. [CrossRef]
6. Kikuta, H.; Toyota, H.; Yu, W. Optical Elements with Subwavelength Structured Surfaces. *Opt. Rev.* **2003**, *10*, 63–73. [CrossRef]
7. Papadopoulos, A.; Skoulas, E.; Mimidis, A.; Perrakis, G.; Kenanakis, G.; Tsibidis, G.D.; Stratakis, E. Biomimetic Omnidirectional Antireflective Glass via Direct Ultrafast Laser Nanostructuring. *Adv. Mater.* **2019**, *31*, e1901123. [CrossRef] [PubMed]
8. Piontek, M.C.; Herrmann, T.; L'Huillier, J.A. Selective glass surface modification with picosecond laser pulses for spatially resolved gloss reduction. *J. Laser Appl.* **2017**, *29*, 22507. [CrossRef]
9. Hauschwitz, P.; Stoklasa, B.; Kuchařík, J.; TurČiČová, H.; Písařík, M.; Brajer, J.; Rostohar, D.; Mocek, T.; Duda, M.; Lucianetti, A. Micromachining of Invar with 784 Beams Using 1.3 ps Laser Source at 515 nm. *Materials* **2020**, *13*, 2962. [CrossRef] [PubMed]
10. Bonse, J.; Kruger, J.; Höhm, S.; Rosenfeld, A. Femtosecond laser-induced periodic surface structures. *J. Laser Appl.* **2012**, *24*, 42006. [CrossRef]
11. Ang, L.K.; Lau, Y.Y.; Gilgenbach, R.M.; Spindler, H.L. Analysis of laser absorption on a rough metal surface. *Appl. Phys. Lett.* **1997**, *70*, 696–698. [CrossRef]
12. Granqvist, C.G.; Hunderi, O. Optical properties of ultrafine gold particles. *Phys. Rev. B* **1977**, *16*, 3513–3534. [CrossRef]
13. Korol'Kov, V.P.; Ionin, A.A.; Kudryashov, S.I.; Seleznev, L.V.; Sinitsyn, D.V.; Samsonov, R.V.; Maslii, A.I.; Medvedev, A.Z.; Gol'Denberg, B.G. Surface nanostructuring of Ni/Cu foilsby femtosecond laser pulses. *Quantum Electron.* **2011**, *41*, 387–392. [CrossRef]

micromachines

Article

Cavity-BOX SOI: Advanced Silicon Substrate with Pre-Patterned BOX for Monolithic MEMS Fabrication

Marta Maria Kluba [1], Jian Li [1,*], Katja Parkkinen [2], Marcus Louwerse [3], Jaap Snijder [3] and Ronald Dekker [1,4,*]

1 The Electronic Components, Technology and Materials (ECTM) Group, Delft University of Technology, Mekelweg 5, 2628 CD Delft, The Netherlands; M.M.Kluba@tudelft.nl
2 Research and Development, Technology, Okmetic Oy, Piitie 2, FI-01510 Vantaa, Finland; Katja.Parkkinen@okmetic.com
3 MEMS & Micro Devices, Philips, High Tech Campus 4, 5656 AE Eindhoven, The Netherlands; Marcus.Louwerse@philips.com (M.L.); Jaap.Snijder@philips.com (J.S.)
4 Philips Research, High Tech Campus 34, 5656 AE Eindhoven, The Netherlands
* Correspondence: J.Li-10@tudelft.nl (J.L.); R.Dekker@tudelft.nl (R.D.)

Abstract: Several Silicon on Insulator (SOI) wafer manufacturers are now offering products with customer-defined cavities etched in the handle wafer, which significantly simplifies the fabrication of MEMS devices such as pressure sensors. This paper presents a novel cavity buried oxide (BOX) SOI substrate (cavity-BOX) that contains a patterned BOX layer. The patterned BOX can form a buried microchannels network, or serve as a stop layer and a buried hard-etch mask, to accurately pattern the device layer while etching it from the backside of the wafer using the cleanroom microfabrication compatible tools and methods. The use of the cavity-BOX as a buried hard-etch mask is demonstrated by applying it for the fabrication of a deep brain stimulation (DBS) demonstrator. The demonstrator consists of a large flexible area and precisely defined 80 µm-thick silicon islands wrapped into a 1.4 mm diameter cylinder. With cavity-BOX, the process of thinning and separating the silicon islands was largely simplified and became more robust. This test case illustrates how cavity-BOX wafers can advance the fabrication of various MEMS devices, especially those with complex geometry and added functionality, by enabling more design freedom and easing the optimization of the fabrication process.

Keywords: SOI substrate; cavity-SOI; cavity-BOX; patterned BOX; buried hard-etch mask; flex to rigid (F2R); MEMS; miniaturization; DBS; foldable devices

Citation: Kluba, M.M.; Li, J.; Parkkinen, K.; Louwerse, M.; Snijder, J.; Dekker, R. Cavity-BOX SOI: Advanced Silicon Substrate with Pre-Patterned BOX for Monolithic MEMS Fabrication. *Micromachines* **2021**, *12*, 414. https://doi.org/10.3390/mi12040414

Academic Editor: Nikolaos Tapoglou

Received: 11 March 2021
Accepted: 6 April 2021
Published: 8 April 2021

1. Introduction

Standard SOI substrates were originally developed to enable perfect dielectric isolation in electronic devices. Nowadays, SOI wafers have also become an important substrate material for the fabrication of MEMS devices. SOI wafers consist of a handle wafer that provides mechanical strength during the fabrication process, a device layer in/on which the devices are fabricated, and a buried oxide (BOX) layer that separates the device layer from the handle wafer (see Figure 1a). Apart from electrical isolation, the BOX layer also allows for the fabrication of MEMS devices with a well-defined device layer thickness, and it can serve as a release layer in floating structures. SOI wafers are used in a wide range of applications, such as pressure sensors [1], resonators and inertial sensors [2], microchannels [3], and miniaturization of microfabricated medical devices [4,5].

Cavity-SOI substrate is a substrate that has been derived from the silicon-on-insulator-family [6]. It is a customized SOI substrate whereby the manufacturer of the SOI substrates has integrated customer-defined buried cavities in the silicon handle wafer (see Figure 1b). It has been demonstrated that customized SOI substrates with prefabricated cavities can significantly simplify the fabrication process of complicated MEMS devices, such as pressure or inertial sensors [2,7–10]. The cavity-SOI substrate allows for eliminating the cumbersome step of etching cavities in the handle layer later in the process flow, and permits for pre-patterning of complex cavities systems. However, the range of the cavity dimensions in the cavity-SOI wafers is limited. On the one hand, very large cavities that are not supported with pillars would make the wafer fragile and could lead to wafer or device layer deformation. On the other hand, very small buried structures are out of cavity-SOI scope, due to the low alignment precision of the prefabricated cavities with the structures fabricated later on the device layer.

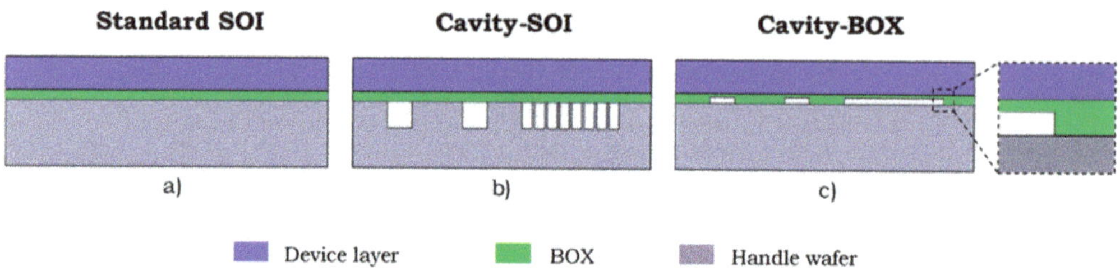

Figure 1. A comparison of three substrate architectures: (**a**) Standard SOI substrate; (**b**) Cavity-SOI substrate with cavities in handle wafer; (**c**) Cavity-BOX substrate with the pre-patterned buried oxide layer.

Cavity-BOX is an advanced substrate with custom-defined cavities etched in the buried oxide (see Figure 1c). It is the newest member of the SOI substrate family. Its exemplar preparation process and application are presented in this paper. The cavities can be formed by etching through the complete thickness of the BOX or by partially etching the BOX to create a hard-etch mask with a step. In cavity-BOX substrates, only the thin layer of buried silicon oxide is patterned, which enables almost unlimited design of the cavities without weakening the mechanical properties of the wafer. The high-precision alignment (<500 nm) of the prefabricated cavities with the structures fabricated later, on the device layer, is ensured by applying a newly developed marker transferring strategy. The method uses a set of primary alignment markers, located on the SOI wafer terrace, that are transferred onto the device layer using front-to-front alignment [11]. The patterned BOX can serve as a stop layer during the device layer thinning, and it can be used as a hard mask during the device layer patterning from the backside. The high resolution of the deep reactive-ion etching (DRIE) process is maintained by bringing the hard mask, formed by the patterned BOX layer, directly to the device layer. This allows for a precise definition of micron-sized cavities in the device layer and, simultaneously, enables patterning centimeter-sized structures in the device layer. This paper presents the novelty of cavity-BOX substrates and illustrates how such a substrate can improve the fabrication of various MEMS devices. The use of the cavity-BOX as a stop layer and a buried hard-etch mask is demonstrated by applying it for the fabrication of a deep brain stimulation (DBS) demonstrator. First, the design and standard fabrication process of such a DBS device are presented and compared with the process that uses the cavity-BOX to show how the cavity-BOX substrate can enable more design freedom and simplify the fabrication process. Next, the preparation of the customized cavity-BOX substrate and the fabrication process of the DBS demonstrator are described. The DBS demonstrator is a mechanical structure composed of only silicon islands connected with a polymer-based flexible film. Finally, the DBS demonstrator fabrication results are presented and discussed.

2. Deep Brain Stimulation (DBS) Probe-Process and Design

The advanced SOI substrate with cavities in the BOX can significantly simplify processes, such as microfabrication of highly integrated foldable devices [4,5] or 3-dimensional circuit integration using TSVs, in the device layer [12]. An example of the cavity-BOX application is a monolithic fabrication process of a foldable deep brain stimulation (DBS) device (see Figure 2 [5]).

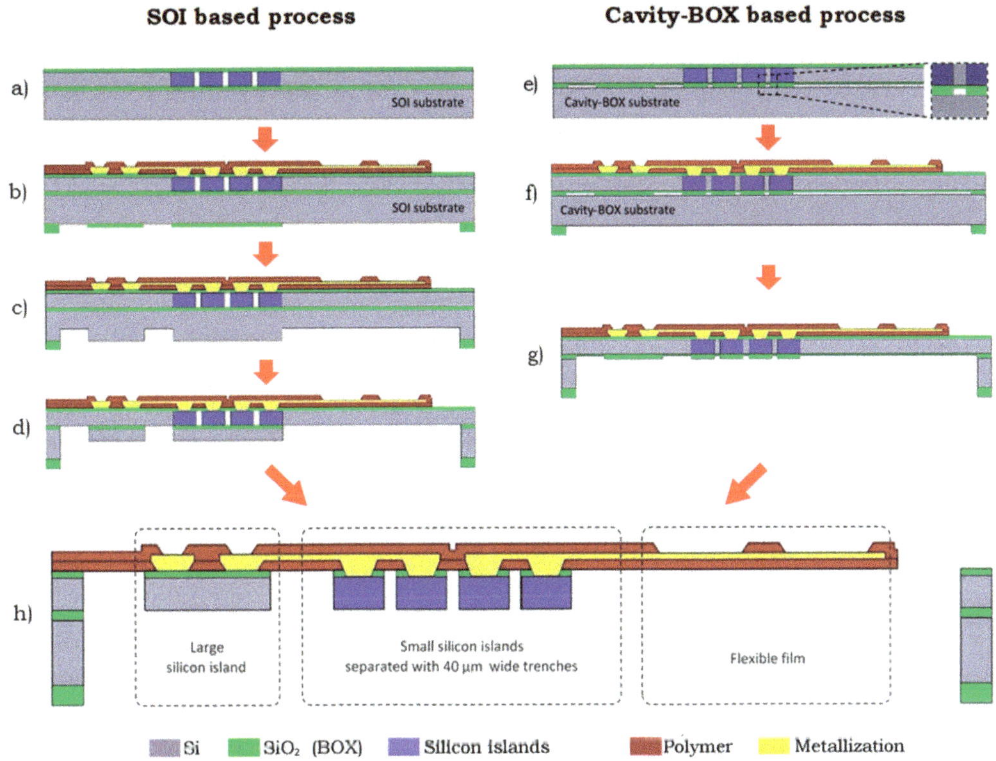

Figure 2. Simplified manufacturing process flow diagrams of semiflexible DBS device using the trench-based F2R technology versus using cavity-BOX substrate. Left (**a–d**): The SOI based process with sealed trenches on the front side of the wafer and a two-step backside etch process. Right (**e–g**): The cavity-BOX based process using patterned BOX as an etch-stop layer and hard-etch mask. Bottom (**h**): Finished device.

Initially, standard SOI substrate and trench-based F2R technology were employed to accomplish the monolithic fabrication of a device where small 80 μm-thick silicon islands, separated with 40 μm-wide trenches, could coexist with a millimeter-sized flexible area etched from the backside of the wafer [5,13]. Due to the resolution limitations of the backside DRIE process, the precise separation of the small (210 × 2070 μm) silicon islands cannot be achieved. Therefore, the HAR trenches were etched in the device layer from the front side of the wafer to separate the silicon islands, and, subsequently, sealed with a silicon dioxide membrane (Figure 2a) to enable further wafer processing (Figure 2b). However, the trench etching and sealing processes require precise optimization, and they have very tight process windows. Moreover, failures of the fragile SiO_2 membrane can severely hamper the follow-up processes. Employing the cavity-BOX substrate with patterned buried oxide (Figure 2e) allows for a more robust process by maintaining the device layer intact until the very end of the front side processing (Figure 2f). All the structures are later released by DRIE etching from the backside using the cavity-BOX as a hard mask.

After the front side processing is finished, the DRIE etching is applied from the backside of the wafer for thinning down and releasing the flexible structures. In the standard SOI process, this is realized by multiple steps of alternating silicon dioxide etch and silicon etch through a two-step hard-etch mask located on the backside of the handle wafer (Figure 2c,d,h). The buried oxide layer of the standard SOI wafer serves as an etch stop layer that defines the device thickness. This approach is cumbersome and heavily relies on the uniformity of each dry etching step. The cavity-BOX can significantly simplify the process by bringing it down to just three steps. First, the device is thinned down to 80 µm using a simple hard-etch mask patterned on the backside of the handle wafer, and the BOX as an etch stop layer that balances the DRIE uniformity of the handle wafer etching (Figure 2g). Secondly, the exposed cavity-BOX with a step mask is thinned down to form the hard-etch mask. Finally, the hard-mask formed in the BOX is used to separate the 80 µm-thick silicon islands with the 40 µm-wide trenches, and simultaneously release the flexible film in the final DRIE step (Figure 2h). The high resolution of the DRIE process and coexistence of the structures with a wide range of dimensions is ensured by bringing the hard mask–patterned BOX directly to the device layer, rather than optimizing the DRIE process to its extreme.

The application of the cavity-BOX substrate is demonstrated using a simplified fabrication process of an 18 mm long Deep Brain Stimulation (DBS) probe (see Figure 3). The highly integrated DBS tip was designed to accommodate 40 circular electrodes on a semi-film substrate. The small silicon islands can contain prefabricated decoupling capacitors, and the large silicon island permits for wire bonding and back-end integration of application specific integrated circuits (ASICs) inside the probe (e.g., flip-chip). All the structures are connected with flexible interconnects. This enables the activation of each electrode individually, using only a couple of power and signal wires reaching out of the probe. As a result of the semiflexible structure, consisting of multiple silicon islands connected with a flexible film, the device can be folded to a 1.4 mm diameter cylinder (Figure 4).

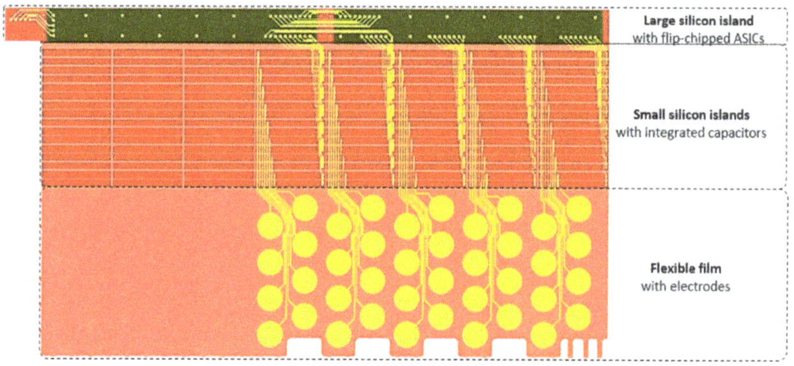

Figure 3. The 2D representation of the 40-electrode DBS design with integrated electronic components. The large silicon island (18 mm × 1 mm) can contain flip-chipped ASICs. The small silicon islands (210 µm × 2070 µm) are separated with 40 µm wide trenches and can accommodate prefabricated decoupling capacitors. The flexible film contains 40 flexible electrodes.

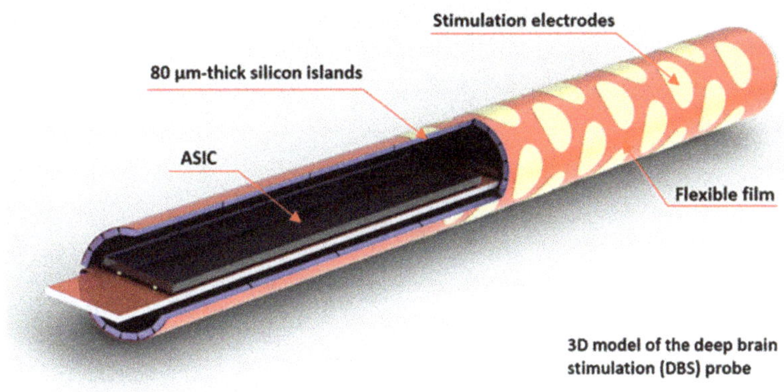

Figure 4. 3D model of the 40-electrode DBS probe (ø 1.4 mm diameter) with integrated capacitors and ASICs inside the probe's tip.

3. Fabrication

The DBS demonstrator was fabricated to illustrate the advances resulting from applying the cavity-BOX to the process. The DBS demonstrator presented here is a semiflexible mechanical structure, composed of silicon island and polymer-based flexible film, without interconnects or integrated electronic components.

The fabrication of the DBS demonstrator can be separated into two parts: the cavity-BOX SOI substrate preparation and the DBS demonstrator fabrication. The main technical challenge arises from the fact that the submicron (less than 1 µm) alignment accuracy between the buried cavity-BOX mask and the structures on top of the device layer must be guaranteed. To overcome that problem, the alignment marker transferring strategy, proposed and developed by C. Mountain et al. [11], was applied to ensure high precision alignment of the structures. Another goal was to demonstrate the functionality of the pre-patterned BOX mask.

3.1. *Cavity-BOX Preparation*

A schematic process flow of cavity-BOX substrate preparation is presented in Figure 5. A 380 µm-thick 6-inch double side polished (DSP) handle wafer was used as a starting material. First, two 140 nm deep ASML markers were patterned into the silicon substrate, 1.2 mm away from the left and right edge of the wafer. Next, 1 µm of high-quality thermal SiO_2 layer for wafer bonding was grown on both sides of the wafer. The customized cavity-BOX pattern was aligned with the markers, and dry etched into the SiO_2 layer, landing on the silicon (Figure 5a). The handle wafer was subsequently fusion bonded with the device layer, which also had a 500 nm-thick layer of thermal oxide. The fusion bonding was carried out under vacuum at room temperature. The two oxide layers were bonded and merged into the cavity-BOX with the pre-patterned step oxide mask (Figure 5b). Finally, the device layer was thinned down to 80 µm, and a terrace with a width of 4 mm was created by a combination of edge trimming and wet etching (Figure 5c). The alignment markers on the handle wafer were revealed during that process.

The 4 mm terrace width was chosen to keep the edge of the device layer as far as possible from the alignment markers on the handle wafer, in case of any possible optical interference during the marker transferring processes. The cavity-BOX substrate is ready after the terracing process. It contained a patterned step buried oxide layer (1.5 µm at its full thickness, 500 nm at its step thickness), an 80 µm-thick device layer, and 1 µm thermal oxide on the backside of the wafer.

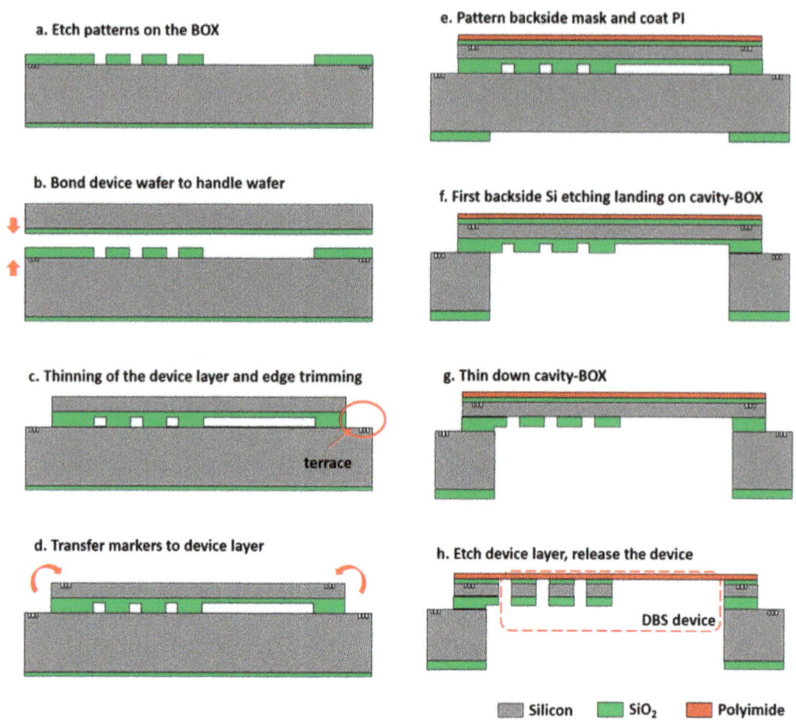

Figure 5. Cross-section drawings of the cavity-BOX substrate preparation and DBS demonstrator fabrication. (**a**) Positioning markers at 1.2 mm to wafer edge and etching patterns in the BOX on top of the 380 μm handle wafer. (**b**) Fusion bonding of the device wafer to the handle wafer with patterned BOX (in vacuum and room temperature. (**c**) Thinning of device layer to 80 μm and edge trimming to create a 4 mm width terrace. (**d**) Transfer markers from the terrace to the device layer. (**e**) Patterning backside etching mask and coating wafer with polyimide, using silicon oxide as an adhesive layer. (**f**) Etching handle wafer from wafer backside and landing on the cavity-BOX. (**g**) Thinning down the cavity-BOX and exposing the buried oxide mask for device layer etching. (**h**) Etching device layer and silicon oxide adhesion layer, landing on polyimide.

3.2. DBS Demonstrator Fabrication

To continue with the demonstrator fabrication, the handle wafer markers were first transferred to the device layer, placed 10 mm from the wafer edge, and etched 140 nm deep into the silicon (Figure 5d). A 2 μm-thick PECVD SiO_2 layer was then deposited on top of the original 1 μm-thick thermal oxide layer on the backside of the wafer and patterned into the silicon DRIE etching mask. Next, a 500 nm-thick PECVD SiO_2 layer was deposited on the front side of the wafer as an adhesion layer. Subsequently, a 3 μm-thick polyimide layer (PI2610 Microsystems) was coated on top of the SiO_2 adhesion layer (Figure 5e) and cured. After that, the silicon DRIE etching step was applied from the backside of the wafer to remove the silicon substrate underneath the cavity-BOX landing on the step oxide mask (Figure 5f). The step oxide mask enabled the silicon over-etch to balance the etching nonuniformity across the wafer. An overall SiO_2 dry etching was subsequently applied to thin down the step oxide mask in the cavity-BOX layer until the pre-patterned oxide mask was opened through to the device layer (Figure 5g). Finally, the 80 μm device layer and the 500 nm SiO_2 layer used for polyimide adhesion were dry-etched, landing on the polyimide layer (Figure 5h). After the etching of the device layer, all the silicon islands were separated, and they were connected with the flexible polyimide film. The finished demonstrator was suspended in a silicon wafer frame through polyimide tabs.

4. Results and Discussion

Both the substrate preparation and the demonstrator fabrication were straightforward. The alignment markers on the terrace edge were successfully detected and transferred to the device layer (Figure 5d). However, there are several other critical steps in the process. The bonding result of the cavity-BOX while preparing for the substrate directly affects its functionality. The terrace width caused an unexpected compatibility problem. Moreover, the backside DRIE etching defines the final device. Here we discuss the aforementioned issues and present the fabricated and assembled DBS demonstrator.

4.1. Preparation of the Cavity-BOX

The cavity-BOX is formed by the bonding process of two oxide layers: 1 μm of pre-patterned SiO_2 layer from the handle wafer and a 500 nm-thick SiO_2 layer from the device layer, with the accuracy of 300 to 500 nm (Figure 5b). The potential concerns of the bonding process include the following: (1) Bonding failures of the small oxide features from the handle wafer; (2) undesired bonding between the silicon substrate of the handle wafer and the SiO_2 layer from the device layer in the large dimension cavities; and (3) sunken surface of the device layer due to the cavities in the BOX.

After the cavity-BOX substrate preparation, no visible sunken surfaces were observed in the device layer. A scanning acoustics microscopy (SAM) was applied to inspect the bonding quality, as shown in Figure 6. The bonded area has a high transmission to the acoustic waves, which are displayed as dark fields in the picture, while air gaps have high reflections on the acoustic waves and are, therefore, displayed as bright fields. The images indicate that the area without cavities was successfully bonded, while there was no undesired bonding in the large cavity areas. No bonding defects have been observed among the small oxide features in the SAM images. Later, after the backside etching of the handle wafer landing on the cavity-BOX (Figure 5f), the two-step oxide mask was intact, which also confirms the excellent bonding result.

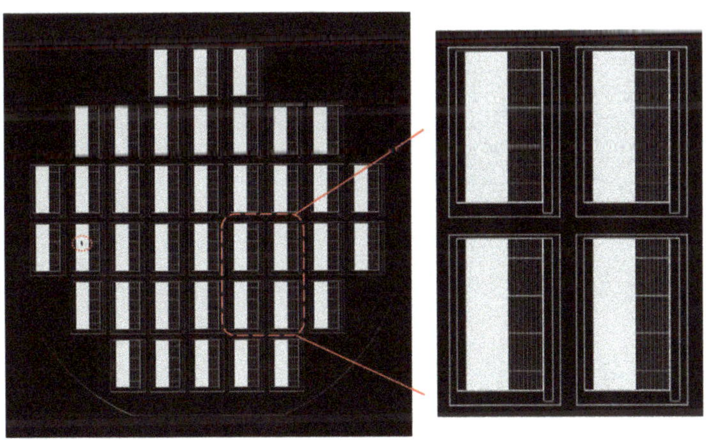

Figure 6. Scanning acoustic microscopy (SAM) to inspect bonding quality. A particle appeared as a black dot on the left side, and a zoomed-in SAM image of four dies.

4.2. Terrace Width

During the backside processing, it appeared that the 4 mm terrace width on the front side of the SOI wafer caused compatibility issues with the PAS5500 ASML wafer stepper and the SPTS Pegasus DRIE etching tool. During the backside lithography process, the wafer stepper needed to handle the wafer from the front side. The positionings of a few vacuum pads on the wafer stepper robot arm were in the 4 mm terrace region, which led to a loss of vacuum. The wide terrace additionally caused helium leakage errors on the chuck of the dry etching tools. To continue processing the 4 mm terrace cavity-BOX SOI wafers, the robot arm of the wafer stepper was explicitly tuned to accept the wafers, and special dry etching recipes with low helium flow were developed for the silicon and SiO_2 etch. As a result of reducing the helium flow in the etching recipes, the wafer temperature could not be well maintained during the process, hence, leading to an increased nonuniform etch rate across the wafer. The etching slop was also enlarged due to the loss of temperature control.

Therefore, the terrace width is preferred to be as small as possible for cleanroom compatibility. In contrast, the terrace edge should be as far as possible from the markers to avoid its interference with the alignment process, requesting a large terrace width. A trade-off test was performed to define the optimal terrace width. Wafers with different terrace widths were tested in the standard 6-inch cleanroom processing line in the PAS5500 ASML wafer stepper, an SPTS Pegasus, ICP, and APS etching tools. The test results indicate that the terrace width of 1.8 to 2.5 mm should provide sufficient compatibility.

4.3. Backside Etching of the Demonstrator

To release the demonstrator and form the semiflexible device structure, several etching steps were performed from the backside of the wafer. These etching steps include the following: (1) The DRIE etching of the handle wafer substrate landing on the cavity-BOX; (2) thinning down the cavity-BOX to form the pre-patterned oxide mask; (3) etching of the device layer and the polyimide adhesion oxide layer using the oxide mask (Figure 5f,h).

Figure 7a depicts the wafer after the first step of the backside Si etching process landing on cavity-BOX. The device layer was protected by the exposed step cavity-BOX. The thickness of the remaining step cavity-BOX after silicon etching was measured using a reflectometer (Nanospec). The thinner part of the step cavity-BOX ranged from 120 nm to 350 nm (originally 500 nm), and the thicker part ranged from 1150 nm to 1380 nm (originally 1500 nm). It can be concluded that the first silicon etching successfully landed on the cavity-BOX, which functions as an etch-stop layer. Figure 7b presents the wafer with the etched device layer after the third backside etching step. The large rectangular opening was transparent, as the etching landed on the polyimide layer coated on the front side of the wafer. All the silicon structures, including the 40 μm-wide trenches between the silicon islands and the gaps that defined the silicon frame, were well fabricated.

(**a**) (**b**)

Figure 7. (**a**) Backside etching result of a silicon substrate, landing on cavity-BOX. (**b**) Etching device layer while using the patterned cavity-BOX as a mask, landing on polyimide.

Figure 8 shows the SEM images of the silicon islands separated with the 40 μm trenches. The top part of the trench between the small silicon islands is 41.4 μm wide, while the bottom part is 54 μm wide. The measurement indicates an under etch of 6.3 μm from each side during the device layer etching, which is not ideal considering the fact that the etching depth is only 80 μm. The large under etching was mainly caused by temperature control issues, which resulted from the lowered helium flow that was necessary to process the wafer on the chuck of the silicon etcher. A better etching profile can be achieved for wafers with compatible terrace width that can be etched using the standard DRIE process without modifying the cooling gas flow.

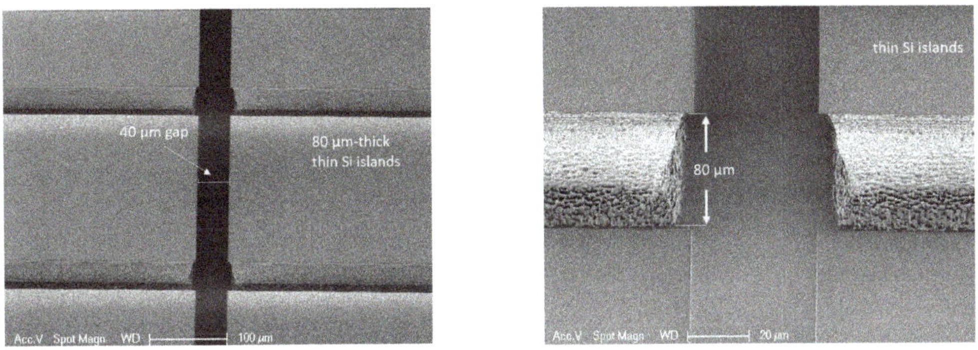

(**a**) (**b**)

Figure 8. (**a**) SEM image of the 40 μm gaps between the silicon islands. (**b**) Zoom-in image of the 80 μm-thick device layer etching profile, with 6,3 μm under-etch from each side.

4.4. Demonstrator Assembly

After fabrication, the demonstrator was taken out of the silicon frame (see Figure 9). The demonstrator consists of one large silicon island that can accommodate ASICs, 128 small silicon islands with 40 μm gaps between them, a large flexible film, and a silicon island for handling. In the final DBS device, the decoupling capacitors are located on each small silicon island, and the electrodes are located on the flexible polyimide film. During the assembly process, the silicon island for the ASICs was attached to a thin metal string with a double-side adhesive Kapton tape. The metal string was slowly rotated, together with the silicon piece, tightening up and wrapping the semiflexible device into a cylindrical probe. In Figure 10, the assembled demonstrator has a length of 18 mm and a diameter of 1.2 mm, which is in line with the DBS design.

Figure 9. DBS demonstrator released after the fabrication. All the 80 μm-thick silicon islands are connected by the flexible film. The small silicon islands are isolated by 40 um-wide gaps.

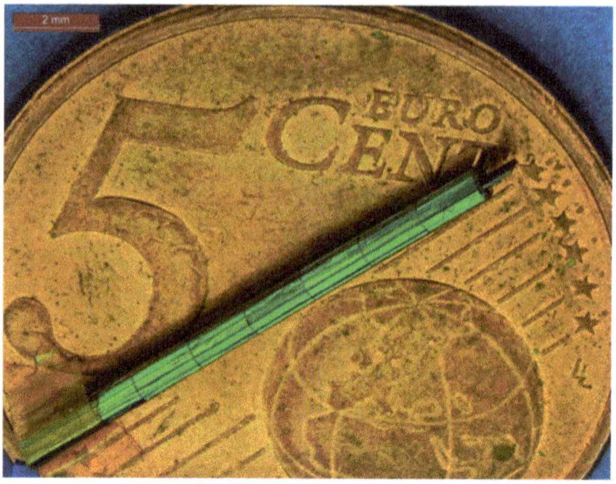

Figure 10. DBS demonstrator wrapped into a cylindrical probe with a length of 18 mm and a diameter of 1.2 μm.

5. Conclusions

Advanced cavity-BOX SOI substrates with a buried oxide mask were developed, prepared, and applied in the process of a semiflexible microfabricated device. The oxide mask was first fabricated on the handle wafer and then bonded with the oxide on the device layer to form the cavity-BOX. The alignment between the structures on the device layer and the buried oxide mask was successfully realized by employing the front-to-front markers transferring strategy. The advanced SOI substrate was implemented for a DBS demonstrator fabrication. The cavity-BOX layer was used as an etch-stop layer and a pre-patterned two-step DRIE mask to etch through the device layer from the backside of the wafer. This two-step oxide layer was successfully used to compensate for etch rate differences during the bulk of the silicon removal. The application of the cavity-BOX substrate to the DBS demonstrator fabrication proved a more robust and significantly simplified process and large design freedom, including the coexistence of high-precision micron-sized features and a large millimeter-sized opening. The semiflexible DBS demonstrator with a length of 18 mm and a diameter of 1.39 mm was successfully fabricated and assembled.

The cavity-BOX SOI has a great potential to simplify the fabrication of various MEMS devices, where device thinning and precise silicon structure separation/definition is needed. The cavity-BOX SOI substrate with a terrace width from 1.8 mm to 2.5 mm is compatible with standard cleanroom equipment for both the front and backside processes. This terrace width also permits applying the high precision (less than 500 nm) alignment strategy of the BOX pattern and the device layer structures. Depending on design requests, the customized cavity-BOX SOI can be provided by the commercial SOI suppliers in large volume, hence, extensively enabling the scaling up of the production.

Author Contributions: Conceptualization, M.M.K. and R.D.; methodology, M.M.K., M.L. and R.D.; validation, M.M.K., J.L., M.L. and K.P.; formal analysis, J.L.; investigation, J.L., J.S. and K.P.; resources, J.L., and K.P.; data curation, J.L. and J.S.; writing—original draft preparation, M.M.K. and J.L.; writing—review and editing, M.M.K., J.L. and R.D.; supervision, M.L. and R.D.; project administration, R.D.; funding acquisition, R.D. All authors have read and agreed to the published version of the manuscript.

Funding: This work started in the European project InForMed (grant no: 2014-2-662155) and now continues as a part of European project POSITION-II (grant no: Ecsel-783132-Position-II-2017-IA).

Acknowledgments: The authors would like to acknowledge Okmetic Oy (Vantaa, Finland) for providing SOI materials and technologies. The cavity-BOX substrate development and preparation were accomplished together with Okmetic Oy (Vantaa, Finland). We would also like to appreciate the effort contributed by the technical staff from Philips MEMS & Micro Devices (MMD) for their respective help in process developments and assembly.

Conflicts of Interest: The authors declare no conflict of interest. The funders had no role in the design of the study; in the collection, analyses, or interpretation of data; in the writing of the manuscript, or in the decision to publish the results.

References

1. San, H.; Zhang, H.; Zhang, Q.; Yu, Y.; Cheng, X. Silicon-glass-based single piezoresistive pressure sensor for harsh environment applications. *J. Micromach. Microeng.* **2013**, *23*, 075020. [CrossRef]
2. Zhang, Y.; Yang, C.; Meng, F.; Liu, G.; Gao, C.; Hao, Y. A monolithic integration multifunctional MEMS sensor based on cavity SOI wafer. In Proceedings of the SENSORS, 2014 IEEE, Valencia, Spain, 2–5 November 2014; pp. 1952–1955.
3. Stoute, R.; Muganda, J.M.; Dahar, S.; Arslan, A.; Henderikx, R.J.M.; van Stiphout, P.C.M.; den Toonder, J.M.J.; Dekker, R. CMOS Compatible Embedded Microchannels. In Proceedings of the 20th International Conference on Miniaturized Systems for Chemistry and Life Sciences, Dublin, Ireland, 9–13 October 2016.
4. Li, J.; Naaborg, J.; Louwerse, M.; Henneken, V.; Eugeni, C.; Dekker, R. Embedded High-Density Trench Capacitors for Smart Catheters. In Proceedings of the 2020 IEEE 8th Electronics System-Integration Technology Conference, ESTC 2020, Online, 15–18 September 2020; pp. 1–5.
5. Kluba, M.; Morana, B.; Savov, A.; van Zeijl, H.; Pandraud, G.; Dekker, R. Wafer-Scale Integration for Semi-Flexible Neural Implant Miniaturization. *Proceedings* **2018**, *2*, 941. [CrossRef]
6. Okmetic. SOI WAFERS. Available online: https://www.okmetic.com/products/product-portfolio/soi-wafers (accessed on 20 December 2018).
7. Liu, C.; Froemel, J.; Chen, J.; Tsukamoto, T.; Tanaka, S. Laterally vibrating MEMS resonant vacuum sensor based on cavity-SOI process for evaluation of wide range of sealed cavity pressure. *Microsyst. Technol.* **2018**. [CrossRef]
8. Wu, G.; Xu, D.; Xiong, B.; Wang, Y. A high-performance bulk mode single crystal silicon microresonator based on a cavity-SOI wafer. *J. Micromech. Microeng.* **2012**, *22*, 25020. [CrossRef]
9. Xu, J.; Zhang, X.; Fernando, S.N.; Chai, K.T.; Gu, Y. AlN-on-SOI platform-based micro-machined hydrophone. *Appl. Phys. Lett.* **2016**, *109*, 32902. [CrossRef]
10. Luoto, H.; Henttinen, K.; Suni, T.; Dekker, J.; Mäkinen, J.; Torkkeli, A. MEMS on cavity-SOI wafers. *Solid State Electron.* **2007**, *51*, 328–332. [CrossRef]
11. Mountain, C.; Kluba, M.; Bergers, L.; Snijder, J.; Dekker, R. Novel method of alignment to buried cavities in cavity-SOI wafers for advanced MEMS devices. *Micro Nano Eng.* **2019**, *5*, 100043. [CrossRef]
12. Shen, W. W.; Chen, K.-N. Three-Dimensional Integrated Circuit (3D IC) Key Technology: Through-Silicon Via (TSV). *Nanoscale Res. Lett.* **2017**, *12*. [CrossRef] [PubMed]
13. Kluba, M.; Arslan, A.; Stoute, R.; Muganda, J.; Dekker, R. Single-Step CMOS Compatible Fabrication of High Aspect Ratio Microchannels Embedded in Silicon. *Proceedings* **2017**, *1*, 291. [CrossRef]

micromachines

MDPI

Article

Numerical Simulation of Effect of Different Initial Morphologies on Melt Hydrodynamics in Laser Polishing of Ti6Al4V

Kai Li [1,2], Zhenyu Zhao [1,*], Houming Zhou [2,*], Hao Zhou [1,2], Jie Yin [1,2], Wei Zhang [1] and Guiyao Zhou [3]

1 School of Intelligent Manufacturing and Equipment, Shenzhen Institute of Information Technology, Shenzhen 518172, China; 201821541904@smail.xtu.edu.cn (K.L.); 201821541865@smail.xtu.edu.cn (H.Z.); 201921001888@smail.xtu.edu.cn (J.Y.); zhang@yeah.net (W.Z.)
2 School of Mechanical Engineering, Xiangtan University, Xiangtan 411105, China
3 Guangdong Provincial Key Laboratory of Nanophotonic Functional Materials and Devices, South China Normal University, Guangzhou 510006, China; Zguiyao@163.com
* Correspondence: yxpzzy01@163.com (Z.Z.); zhouhouming@xtu.edu.cn (H.Z.)

Abstract: As a surface finishing technique for rapid remelting and re-solidification, laser polishing can effectively eliminate the asperities so as to approach the feature size. Nevertheless, the polished surface quality is significantly sensitive to the processing parameters, especially with respect to melt hydrodynamics. In this paper, a transient two-dimensional model was developed to demonstrate the molten flow behavior for different surface morphologies of the Ti6Al4V alloy. It is illustrated that the complex evolution of the melt hydrodynamics involving heat conduction, thermal convection, thermal radiation, melting and solidification during laser polishing. Results show that the uniformity of the distribution of surface peaks and valleys can improve the molten flow stability and obtain better smoothing effect. The high cooling rate of the molten pool resulting in a shortening of the molten lifetime, which prevents the peaks from being removed by capillary and thermocapillary forces. It is revealed that the mechanism of secondary roughness formation on polished surface. Moreover, the double spiral nest Marangoni convection extrudes the molten to the outsides. It results in the formation of expansion and depression, corresponding to nearby the starting position and at the edges of the polished surface. It is further found that the difference between the simulation and experimental depression depths is only about 2 μm. Correspondingly, the errors are approximately 8.3%, 11.3% and 13.3%, corresponding to Models 1, 2 and 3, respectively. The aforementioned results illustrated that the predicted surface profiles agree reasonably well with the experimentally measured surface height data.

Keywords: laser polishing; different surface morphologies; melt hydrodynamics; numerical simulation

Citation: Li, K.; Zhao, Z.; Zhou, H.; Zhou, H.; Yin, J.; Zhang, W.; Zhou, G. Numerical Simulation of Effect of Different Initial Morphologies on Melt Hydrodynamics in Laser Polishing of Ti6Al4V. *Micromachines* **2021**, *12*, 581. https://doi.org/10.3390/mi12050581

Academic Editor: Nikolaos Tapoglou

Received: 6 May 2021
Accepted: 18 May 2021
Published: 20 May 2021

Publisher's Note: MDPI stays neutral with regard to jurisdictional claims in published maps and institutional affiliations.

1. Introduction

Ti6Al4V alloy has been widely used in aerospace [1], chemical industries [2] and medical sector [3] due to its low density [4], high strength [5] and excellent biocompatibility [6]. Therefore, the surface quality, such as surface roughness, of the devices fabricated from Ti64 is posed in a great challenge. At present, the polishing of Ti64 surface is mainly by traditional methods such as mechanical polishing, electrochemical polishing and plasma polishing. It is difficult to polish the complex surface. Laser polishing (LP) as a non-contact processing surface treatment technology, with high polishing efficiency, non-pollution and a high degree of processing flexibility. Additionally, LP not only can effectively surface roughness, but also the polished surface can be secondary strengthening such as hardness [7,8], wear and corrosion resistance [9,10]. Additionally, polishing on Ti6Al4V is a real challenge, which can modify the surface and becoming fragile and Widmanstätten patterns during PAM process [11,12]. This work can prevent the realization of such damage or wrong patterns.

Nonetheless, LP is a complex physical process. Owing to the extremely short interaction duration of the laser with the material, it can reach the millisecond or even microsecond scale. Thereby, it is difficult to observe the liquid metal flow and the evolution of the polished surface topography. Hence, a large number of scholars have predicted surface morphology and melt hydrodynamics by means of FEM models. The concept of surface micro-melting was first introduced by Mai et al. [13]. A hybrid numerical model was developed based on the advantages of variable and fixed domains to calculate the non-linear problem of metal solid-liquid moving boundary. Based on the damped oscillatory behavior generated by the molten surface, Perry et al. first introduced the one-dimensional critical frequency f_{cr} to predict the polishing effect of the polished surface within the spatial frequency domains [14–16]. When the spatial frequency amplitude of the polished surface is greater than the f_{cr}, the surface roughness is significantly decreased. Vadali et al. extended the concept of one-dimensional critical frequency to the two-dimensional plane based on a series of physical equations that can predict the spatial frequency content and surface roughness after polishing [17]. The f_{cr} was accurately obtained by solving the heat conduction differential equation by Ukar et al. [18]. It also predicted the polished surface morphology well. Further, Wang et al. developed a surface prediction model for thermocapillary regime smoothing [19]. They applied the capillary force prediction model proposed by Vadali et al. [17] to predict the spatial spectrum of polished surface by using introduced feature slope and normalized average displacement.

The above simulation process is only a semi-empirical surface prediction model that cannot simulate the evolution of the melt hydrodynamics. Sim et al. developed a numerical model of the axisymmetric thermocapillary flow during laser melting and further analyzed the evolution of the molten free surface with radiation duration [20]. Nevertheless, the solidified surface morphology cannot be determined due to the cooling process is not considered. Ma et al. proposed a two-dimensional (2D) axisymmetric transient model dominated by Marangoni convection. Since the model assumed that the polished surface is an ideal smooth surface [21]. Thus, only Marangoni flow was contained in the molten pool and no capillary flow associated with surface curvature was introduced. In further, Zhang et al. performed a model illustrates the molten flow behavior and the formation of free surface during LP and the contribution of capillary and thermocapillary forces to the melt hydrodynamics were analyzed [22]. The result indicates that capillary forces dominate in the initial stage of melting, mainly eliminating the surface with large curvature. Correspondingly, the thermocapillary force is dominant in the melt development stage and further improves the quality of polished surface. However, there are numerous factors affecting the polishing process. The influence of different initial surface morphologies on the evolution of melt hydrodynamics has been few investigated.

In the present work, a transient 2D model based on a moving heat source is developed by coupling the heat transfer field and the flow field to simulate the evolution of the liquid metal surface from rough to smooth. The mapping law of the melt hydrodynamics with different initial surface features was detailed examined from three aspects: temperature field, velocity field and the evolution of different molten surface profiles. In particular, the underlying cause of the bumps formation during polishing process was analyzed. To verify the numerical model, laser polishing experiments were performed on three different surfaces as well as polished surface were measured for comparison with the predicted surface profile.

2. Numerical Simulation

2.1. Governing Equations

To ensure the accuracy while saving computational costs, the laser polishing model was developed based on the following assumptions.

(a) The property of fluid phase fluid is treated as incompressible Newtonian laminar flow.
(b) The material distribution satisfies continuity and isotropy.
(c) The laser incident energy is considered as the surface heat flux.

(d) Due to the ratio of the density of the liquid Ti6Al4V and the dynamic viscosity argon gas of is large, the influence of gas flow on the free surface evolution can be neglected.

In this paper, the model is based on the energy equation (Equation (1)), momentum equation (Equation (2)) and continuity equation (Equation (3)) as the theoretical guidance [22–26].

$$\rho C_p^*[\frac{\partial T}{\partial t} + \nabla \cdot (\vec{u}T)] - \nabla \cdot (k\nabla T) = 0 \quad (1)$$

$$\rho\left(\frac{\partial \vec{u}}{\partial t} + \left(\vec{u} - \vec{u}_m\right) \cdot \left(\nabla \vec{u}\right)\right) = \nabla \cdot \left(-pI + \mu\left(\nabla \vec{u} + \left(\nabla \vec{u}\right)^T\right)\right) + F_V \quad (2)$$

$$\nabla \cdot \vec{u} = 0 \quad (3)$$

where ρ is the density, t is the laser radiation duration, T is the the variation of surface temperature with t, $\vec{u}$ is the melt velocity, k is the thermal conductivity, $\vec{u}_m$ is the mesh velocity, p is the pressure, I is the identity matrix, μ is the dynamic viscosity, F_V is the body force of buoyancy and gravity of the molten pool [23].

$$F_V = \rho_{ref}g\left(1 - \beta\left(T - T_{ref}\right)\right) \quad (4)$$

where T_{ref} is the reference temperature, ρ_{ref} is the reference density, β is the thermal expansion coefficient and g is the gravity constant.

To balance the energy, the latent heat of melting term released during the solid to liquid phase transition. It is added to the specific heat capacity function as an equivalent heat capacity using a Gaussian function in the simulation [22].

$$C_p^* = C_p + L_m(\frac{df_L}{dt}) \quad (5)$$

where C_p is the specific heat, L_m is the latent heat of melting, the definition of liquid fraction f_L as follow [22–26]

$$f_L = \begin{cases} 0 & T \le T_s \\ \frac{T-T_s}{T_l-T_s} & T_s \le T \le T_l \\ 1 & T_l \le T \end{cases} \quad (6)$$

where T_s is the solid phase temperature, T_l is the liquid phase temperature.

In addition, Table 1 illustrates the thermophysical properties of Ti6Al4V are used during the modeling process.

Table 1. Thermophysical properties of Ti6Al4V [22–27].

Parameter	Nomenclature	Value
Solidus temperature (K)	T_s	1877
Liquidus temperature (K)	T_l	1923
Melting temperature (K)	T_m	1900
Boiling temperature (K)	T_b	3315
Ambient temperature (K)	T_a	298.15
Solidus density (kg m^{-3})	ρ_s	4420
Liquidus density (kg m^{-3})	ρ_l	4000
Dynamic viscosity (Pa s)	μ	0.005
Solidus thermal conductivity (Wm^{-1} K^{-1})	k_s	21
Liquidus thermal conductivity (Wm^{-1} K^{-1})	k_l	29
Solidus specific heat (Jkg^{-1} K^{-1})	$C_{p\text{-}s}$	670

Table 1. *Cont.*

Parameter	Nomenclature	Value
Liquidus specific heat (Jkg^{-1} K^{-1})	$C_{p\text{-}l}$	831
Convective coefficient (Wm^{-2} K^{-1})	h	10
Temperature derivative of surface tension (Nm^{-1} K^{-1})	$\partial\gamma/\partial T$	-2.8×10^{-4}
Latent heat of melting (Jkg^{-1})	L_m	2.86×10^5
Emissivity	ε	0.6
Absorptivity	α_0	0.3

2.2. Model Geometry

Ramos et al. assumed that the free surface of selective laser sintering part is composed of closed hemispherical caps during study SSM mechanism of LP [28–31]. However, the experimental polished surface is more asperities, consisting of various irregular peaks and valleys. In order to establish an accurate geometric model, firstly, a 3D optical profilometer based on white light interferometer technology was used for initial surface profile inspection and results as shown in Figure 1a. Then, due to high frequency noise of 3D optical surface profile will cause the odd occurrence of surface during the geometric modeling process. Thus, a rectangular window function is added to satisfy the Fourier filtering of the surface high frequency features, so as to obtain smooth geometric features closer to the experimental surface. The filtered surface morphology is shown in Figure 1b.

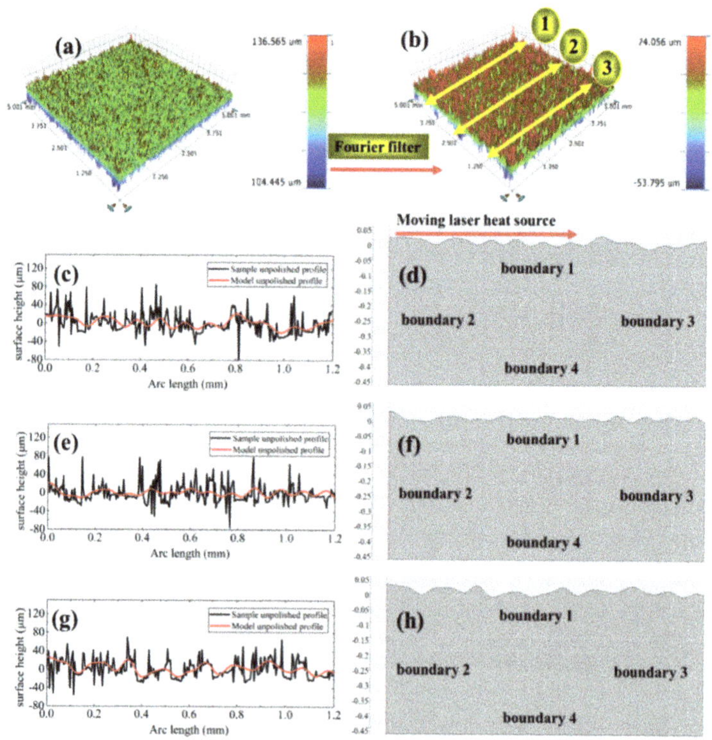

Figure 1. Geometry model of molten pool. (**a**) Optical morphology of initial surface; (**b**) Optical morphology of initial surface after Fourier filtering; (**c**) Surface profile height of Model 1; (**d**) Geometry of Model 1; (**e**) Surface profile height of Model 2; (**f**) Geometry of Model 2; (**g**) Surface profile height of Model 3; (**h**) Geometry of Model 3.

Furthermore, to investigate the effect of different initial surface morphologies on the molten pool characteristics, three different linear surface profiles were extracted from the filtered surface, as shown in Figure 1c,e,g. The black and red lines are, respectively, representing the initial and the filtered linear profile. Then, the filtered surface data is imported into COMSOL by means of the interpolation function to build the geometric model, which is defined as Model 1 (see Figure 1d), Model 2 (see Figure 1f) and Model 3 (see Figure 1h). It is noted that the Model 1 surface is relatively uneven, with a large difference in the height of the surface profile of the peaks and valleys. Especially in the surface with negative curvature morphology near the right edge. The distribution of peaks and valleys on the surface of Model 2 is relatively uniform. The height difference of the surface profile is relatively stable without a large curvature profile. In comparison to Models 1 and 2, the Model 3 has a larger distance between the peaks and valleys, namely, a larger wavelength. Meanwhile, a larger profile height along with a reduced number of peaks and valleys as well as a nonuniform distribution. In summary, different surface topography characteristics will not only affect the absorption of laser heat, but also have an impact on the molten pool flow and the distribution of driving force and further determine the polished surface quality.

2.3. Boundary Conditions

(1) Heat transfer boundary condition

Equation (7) describes the laser beam radiation which induces the thermal convection and surface to surface ambient radiation heat loss whiles the thermal insulation in boundary 4 is expressed in Equation (8) [22–26].

$$-k\nabla T = h(T - T_a) + \varepsilon\sigma\left(T^4 - T_a^4\right) \tag{7}$$

$$\nabla T = 0 \tag{8}$$

where ε is the surface emissivity, σ is the Stefan-Boltzmann constant, T_a is the ambient temperature, h is the convective coefficient.

(2) Momentum boundary condition

Equation (9) describes the velocity field at the boundary which is set as no slip wall in boundary 4 whiles the flow velocity in the r direction which is limited to zero in boundaries 2 and 3 is expressed in Equation (10) [25].

$$u_r = u_z = 0 \tag{9}$$

$$u_r = 0, \ \frac{\partial u_z}{\partial r} = 0 \tag{10}$$

where u_r and u_z is the fluid flow velocity along the r and z directions.

(3) Free surface boundary condition

Equation (11) describes the total stress act on the free surface (boundary 1) of the molten pool. Additionally, the normal and tangential components are expressed in Equations (12) and (13) [22–26].

$$\sigma = \left(\nabla \cdot \vec{n}\right)\gamma\vec{n} - \nabla\gamma \tag{11}$$

$$\sigma_n = \left(\nabla \cdot \vec{n}\right)\gamma\vec{n} = \kappa\gamma\vec{n} \tag{12}$$

$$\sigma_t = \nabla\gamma = \frac{\partial\gamma}{\partial T}\nabla_s T \cdot \vec{t} = \frac{\partial\gamma}{\partial T}\left[\nabla T - \vec{n}\left(\vec{n} \cdot \nabla T\right)\right] \tag{13}$$

where γ is the surface tension coefficient, κ is the surface curvature. $\partial\gamma/\partial T$ is the temperature gradient of surface tension, $\vec{n}$ and $\vec{t}$ are represent the normal and tangential vectors, respectively.

2.4. Laser Moving Heat Source

The model uses a top-hat laser heat source with uniform energy distribution. Accordingly, the stationary and moving heat source are, respectively, expressed by Equations (14)–(17) [32].

$$Q_s = \frac{P}{\pi R_0^2} \tag{14}$$

$$f(r^*) = \begin{cases} 1 & |r^*| \leq r_0 \\ 0 & |r^*| \geq r_0 \end{cases} \tag{15}$$

$$r^* = r - vt - 0.05 \tag{16}$$

$$Q_m = Q_s \times f(r^*) \tag{17}$$

where P is the laser power, r_0 is the spot radius, r is the independent variable in the cylindrical coordinate system, v and t are the laser moving velocity and time, 0.05 (unit: mm) is the starting position of the laser polished surface. Furthermore, the product of piecewise function $f(r^*)$ and Q_s describe the laser energy density acting within the laser beam. In addition, the modeling process parameters and specific boundary conditions set in physical fields are shown in Tables 2 and 3, respectively.

Table 2. Modeling process parameters in LP.

Polishing Parameter (Unit)	Nomenclature	Value
Laser beam radius (mm)	R_0	0.135
Laser power (W)	P	150
Laser scanning speed (mm s^{-1})	v	300
Laser heating duration (ms)	t_h	3.5
Cooling duration (ms)	t_c	0.3

Table 3. Summary of physical field boundary conditions.

Boundary Condition	Boundary (See Figure 1)	Physical Condition
Boundary heat source	1	Laser radiation
Convection	1, 2, 3	Natural convection
Diffuse surface	1, 2, 3	Radiation
Thermal insulation	4	Insulation
Capillary force	1	Weak contribution
Themocapillary force	1	Marangoni effect
Wall	2, 3,4	No slip wall

2.5. Moving Mesh

The model tracks the deformation of the liquid/gas interface by means of moving mesh. Moreover, coupling of Arbitrary Lagrangian-Eulerian (ALE) method with momentum conservation equation expression is shown in Equation (18) [33].

$$u_m \cdot \vec{n} = u_{mat} \cdot \vec{n} \tag{18}$$

where u_m is the moving velocity of the mesh, u_{mat} is the material velocity.

2.6. Mesh and Configurations

In order to calculate the displacement of the free surface accurately, the free surface is hydrodynamically calibrated with a maximum cell size of 0.8 μm and a maximum cell growth rate set to 1.05. Owing to most of the solution area is a solid, a general physical calibration is used in the solution domain with a maximum cell size of 20 μm so as to save computational cost. The maximum cell size is 20 μm, and the maximum cell growth rate is set to 1.1. Most importantly, the moving mesh needs to be Laplace smoothed. If hyper-elasticity or Yeoh smoothing is used, it may lead to singularity and nonconvergence of the top surface mesh [34,35]. The specific mesh parameter for the solution domain as shown in Table 4. The mesh division results are shown in Figure 2. Owing to the different initial morphologies, the number of free triangles in the solution domain varies. The number of mesh cells for models 1, 2 and 3 are 56,496, 56,056 and 57,182, respectively. The average cell quality is above 88% for each model. Meanwhile, the calculation for each model requires about 12 h on a computer equipped with 16.0 GB RAM and four Intel(R) Core (TM) i7-9700K processors at 3.60 GHz CPU speed.

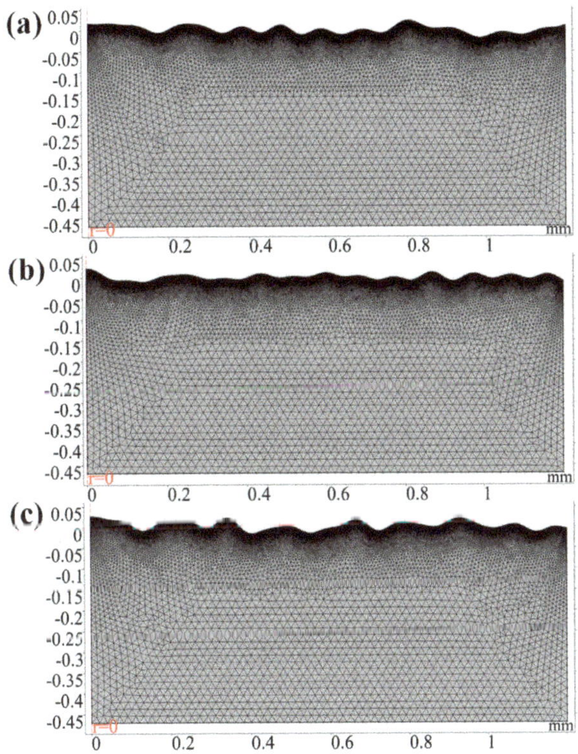

Figure 2. Meshed geometry. (**a**) Model 1; (**b**) Model 2; (**c**) Model 3.

Table 4. Free triangular element size parameters.

Parameter (Unit)	Top Layer	The Rest
Maximum element size (μm)	0.8	20
Minimum element size (μm)	0.002	0.024
Maximum element growth rate	1.05	1.1
Curvature factor	0.2	0.2

3. Experimental Setup and Methods

3.1. Polishing Experimental Setup

The purpose of the experimental study is to verify the established numerical model of the LP melt pool dynamics of Ti6Al4V alloy. The schematic and physical diagrams of the polishing experimental system, as shown in Figure 3. The system consists of laser transmitter, beam expander, dynamic focusing system, 2-axis CNC rotary table and protective gas device [36]. The laser transmitter is a single-mode continuous wave fiber laser (Model: MFSC-1000 W, from Shenzhen Chuangxin Laser Co., Ltd., Shenzhen, China) with adjustable power from 150 W to 1000 W [37]. Furthermore, the laser beam passes through the beam expander (Model: 2-8-355-200 m, from Nanjing Wavelength Opto Electronic Pte., Ltd., Nanjing, China) with a top-hat energy distribution, corresponding to a maximum beam quality M^2 of less than 1.3 [37]. The dynamic focusing system (Model: SDL-F20PRO-3, from Suzhou FEELTEK Laser Technology Co., Ltd., Suzhou, China) with a maximum polishing area of 600 × 600 mm^2 can obtain a laser beam with a maximum scanning speed of 4000 mm/s, with the focal point of the laser beam generated at 720 mm from the polished surface [36]. Additionally, argon gas, with a purity of 99.99%, is used as a shielding gas to fill the processing chamber to prevent surface oxidation during the polishing process.

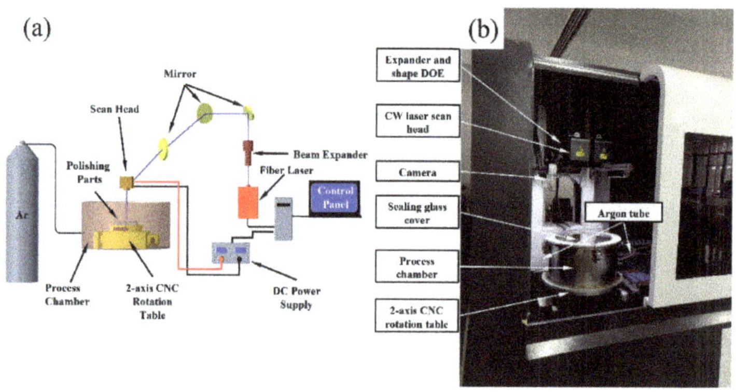

Figure 3. Experimental setup and device of the polishing on Ti6Al4V. (**a**) Experiment device; (**b**) Principle of experiment device.

3.2. Experimental Methods

The material used for the polishing experiments was Ti6Al4V alloy, whose main chemical element compositions are shown in Table 5 [22]. Prior to polishing experiment, the initial surface is cleaned with anhydrous ethanol to prevent dust and other impurities from reducing the polishing quality. Furthermore, in order to ensure that the simulation model is consistent with the experiment, the polishing was performed at the laser focus when the power of 150 W, the scanning speed is 300 mm/s. Finally, the surface morphology of the single-line polished tracks for Models 1, 2 and 3 (see Figure 12a,c,e) was obtained by the 3D optical profilometer (Model: ContourGT-X, from Bruker Nano, Inc., Tucson, AR, USA) based on white light interferometer technology.

Table 5. Main chemical composition of Ti6Al4V (wt %).

Ti	Al	V	C	Fe	O	N
Balance	5.50–6.75	3.50–4.50	0.08	0.30	0.20	0.05

4. Results and Discussion

4.1. Molten Flow Behavior of Model 1

Figure 4 demonstrates evolution of molten pool morphology and the distribution of temperature and velocity fields for Model 1. The polishing was performed at the laser focus when the power of 150 W, the scanning speed is 300 mm/s. The black line is the isotherm of the radiation region, the upper isotherm is the liquidus temperature, the lower is the solidus temperature. The region between the two isotherms is defined as mushy zone [38–40]. From the simulation results, it can be seen that the polished surface starts to melt and forms a shallow molten pool of 9 μm on the surface at about 0.3 ms. After which the depth of the molten pool expands continuously with the increase of laser radiation duration until it reaches about 60 μm at 3.5 ms. Furthermore, it is worth noting that the overall molten pool width is also gradually increasing, but at 2.5 ms, it can be found that the extent of the paste zone is smaller compared to 0.5~0.2 ms. It is due to the high cooling rate of the molten pool resulting in a larger area of molten pool solidification at the tail of the spot than the expansion area of the melt front. Moreover, the depth and width of the molten pool continued to increase gradually in the heating duration range from 2.5 ms to 3.5 ms. It reaches a maximum of 58 μm and 435 μm at 3.5 ms. When the laser radiation shut off, the depth and width of the molten pool rapidly decreased to 55 μm and 330 μm after a cooling time of 0.05 ms. The molten pool was completely solidified until 3.8 ms.

4.2. Analysis of Temperature Field

From the temperature field distribution, it can be seen that the highest temperature of the polished surface always appears the position near the trailing edge of the laser beam. When the surface temperature above the melting temperature at 0.3 ms, the spot position moves from left to right with the heating duration. Owing to the heat conduction, the surface accumulates much heat, the temperature of the area radiated by the spot gradually increases at 0.3~1.5 ms. Then it reaches the maximum value of about 2760 K. In further, as the laser radiation duration from 1.5 ms to 3.5 ms, surface temperature decreases and is accompanied by small fluctuations. It is due to the expansion of the molten pool area and the increase of the melt flow rate generating larger heat convection and radiation. It results in losing some of the heat, thereby affecting the surface temperature distribution. Simultaneously, it can be observed that the heat affected area of the polished region expands with increasing radiation duration. During the cooling period, the molten pool can no longer obtain heat flux from the laser radiation. With the transfer of heat from the high temperature region of the molten pool to the low temperature region, the surface temperature plummets below the solidus temperature within 0.3 ms after the laser heating is stopped.

4.3. Analysis of Velocity Field

Furthermore, it can be seen from the velocity field distribution that the maximum flow velocity always occurs near the solid and liquid phase regions on both sides of the pool. At 0.3 ms, the temperature of the surface peaks reaches above the melting temperature (1900 K) and begins to melt. At this time, the material dynamic viscosity decreases, the capillary force and thermocapillary overcomes the viscous stress and drives the melt to flow from the peaks to the valleys (see Figure 5). It is worth noting that the thermocapillary forces dominate the molten pool in the region of $12 < r < 93$ μm as tangential flow along the surface. Correspondingly, in the region of $93 < r < 150$ μm, the capillary forces dominate the melt with large curvature normal to the surface along the peaks towards the valleys. In addition, due to the negative surface tension temperature coefficient of Ti6Al4V alloy, the higher the temperature of the polished surface, the lower the surface tension in the area. Thus, the thermocapillary forces cause the melt to flow from the center of the beam (low surface tension) to the edge of the molten pool (high surface tension). Meanwhile, the temperature gradient near the edge of the trailing molten pool is greater than that at the center of the beam and at the melt front. Therefore, the maximum velocity of the molten pool always occurs at the edge of the molten pool as well as the maximum velocity of the molten pool is 0.35 m/s. At the beginning of LP, namely in the heating range of 0.3 ms to 1.5 ms, the temperature gradient at the trailing edge of the molten pool increases as the surface temperature rises. It results in more pronounced Marangoni convection or thermocapillary forces. Accordingly, the molten pool velocity increases from 0.35 m/s to 1.97 m/s. From 1.5 ms to 3.5 ms, the molten pool flow velocity does not vary much, reaching a maximum flow velocity of about 2.25 m/s at the 3 ms moment. When the laser radiation stops, the temperature gradient in the molten pool area decreases due to the sharp drop in surface temperature. Moreover, the dynamic viscosity of the liquid metal gradually transitions to high viscosity, namely, the viscosity within the molten pool is enhanced. It causes the melt flow rate to drop to 0.44 m/s until the molten pool is completely solidified at 3.8 ms. At this point, the very high dynamic viscosity limits the molten pool velocity to close to zero.

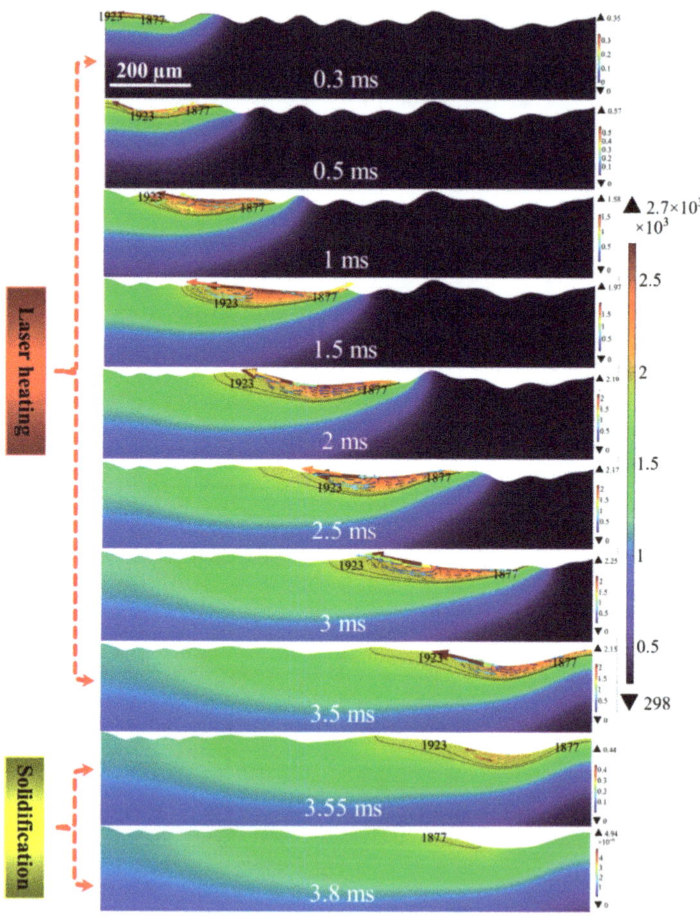

Figure 4. Evolution of molten pool morphology and the distribution of temperature field (color surface contour, unit: K) and velocity field (colored arrow plots, unit: m/s) of Model 1.

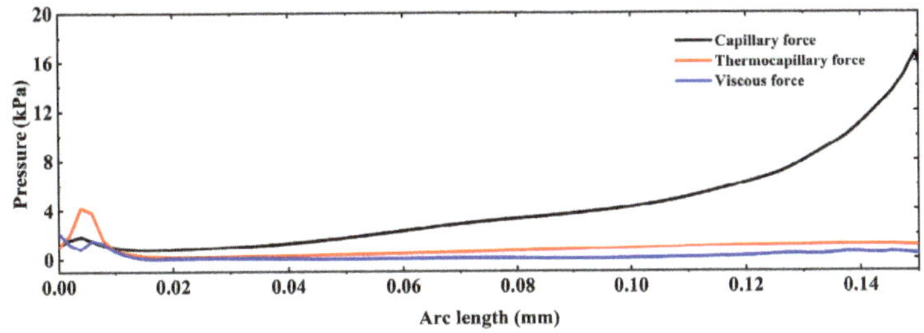

Figure 5. The dominant of capillary and thermocapillary forces as well as viscosity distribution at 0.3 ms heating duration.

4.4. Analysis of Free Surface Evolution

The evolution of the free surface morphology of the molten pool indicates that the shallow part of the surface starts to melt at a polishing duration of 0.3 ms. Due to the small melting area, the molten pool morphology did not change significantly. At 0.5 ms, the trailing melt of the molten pool flows from right to left by Marangoni convection and form half peaks. As the laser beam continues to move to the right, the peaks begin to melt and the melt flows from the peaks to the valleys under the combined effect of normal capillary force, tangential thermocapillary force and body force. At 3 ms, it can be found that the height of the surface profile of the polished peaks is significantly reduced compared to the initial moment and the number of peaks is reduced. It is worth noting that at a polishing time of 3.5 ms, the molten pool surface forms depressions larger than the spot diameter at the edges. It is due to the Marangoni convection formed by the temperature gradient between the center and the edge of the beam causing the melt to flow from the center to the outside. Additionally, the high cooling rate of the molten pool prevented the capillary force from having time to smooth the depressions normally, and a valley with a depression depth of 26 μm was formed on the surface at 3.8 ms.

4.5. Analysis of Secondary Surface Bumps Formed

In addition, it can be found that the contour height of the surface peaks has been reduced, but it has not yet reached a smooth effect. In order to analyze the causes of this phenomenon, the evolution duration of the molten pool morphology was further refined. Figure 6 represents the evolution of the molten pool morphology for a polishing time of 0.7~1 ms for Model 1. At 0.7~0.8 ms, a double swirl Marangoni convection can be found on both sides of the molten pool. At the same time, the center of the molten pool is swollen by the buoyancy force, causing a local bump on the surface. At 0.91~0.92 ms, it can be observed that the left side of the molten pool is dominated by capillary forces that make the molten projections flow along the surface normal to the melt. Thus, it results in the melt to be extruded on both sides. Further, a Marangoni convection acting tangentially along the surface was formed at 0.95 ms in the trailing molten pool again. Simultaneously, due to the surface temperature approaching the solid phase temperature, the melt dynamic viscosity increases as well as the capillary force is no longer able to normalize the bulge profile. Thus, the surface is formed a bulge profile at 1 ms. Furthermore, to eliminate or reduce the surface roughness caused by peaks, the laser power can be increased, or the scanning speed can be reduced to increase the survival life of the molten pool. It allows for adequate smoothing of the polished surface by capillary and thermocapillary forces.

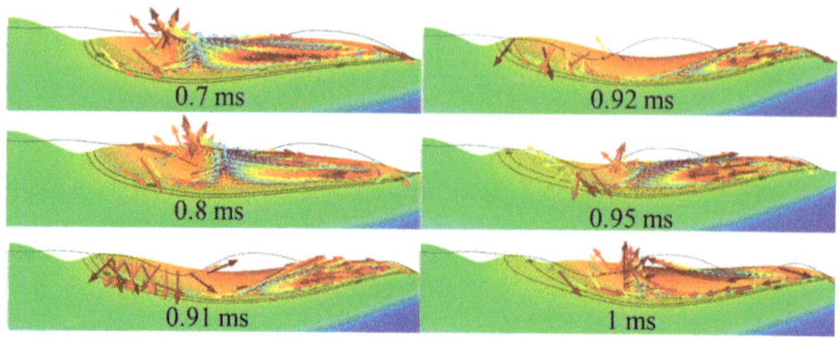

Figure 6. Evolution of local molten pool morphology and the distribution of temperature field (color surface contour, unit: K) and velocity field (colored arrow plots, unit: m/s) of Model 1.

4.6. Molten Flow Behavior of Models 2 and 3

The evolution of local molten pool morphology and the distribution of temperature field and velocity field for Models 2 and 3 are shown in Figures 7 and 8, respectively. Since this model uses a fixed absorption rate, Models 2 and 3 have similar general patterns in the temperature field compared to Model 1. Namely, the temperature of the polishing area is increased with the increase of radiation time, the highest temperature are in the trailing edge of the beam and stable at about 2700 K. The results show that different initial surface morphologies have a little effect on the temperature field. In addition, for the same input laser energy density, the molten pool temperature field distribution depends mainly on the absorption rate of the laser by the material surface. In terms of velocity field, the flow direction of the molten pool is from the center of the spot to the edge, and the maximum velocity occurs near the melt front.

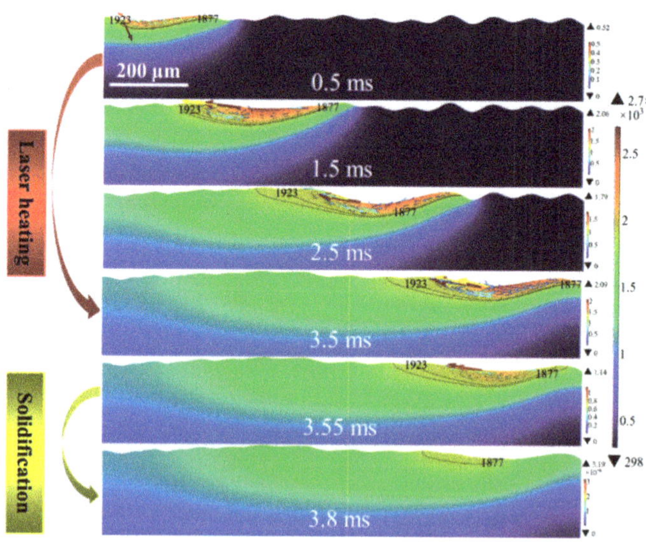

Figure 7. Evolution of molten pool morphology and the distribution of temperature field (color surface contour, unit: K) and velocity field (colored arrow plots, unit: m/s) of Model 2.

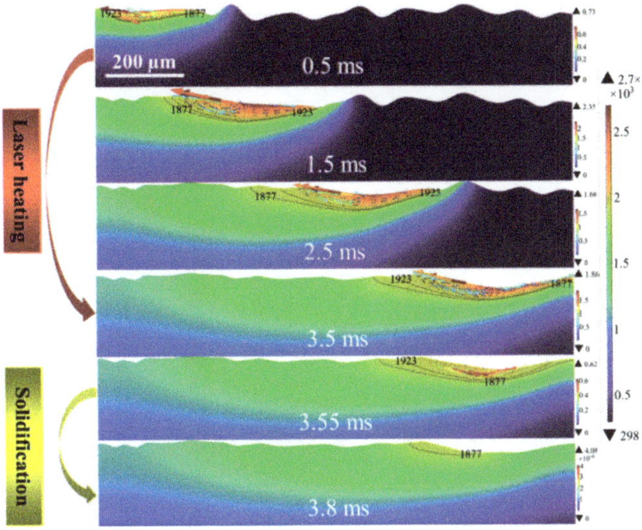

Figure 8. Evolution of molten pool morphology and the distribution of temperature field (color surface contour, unit: K) and velocity field (colored arrow plots, unit: m/s) of Model 3.

4.7. Evolution of Melt Hydrodynamics for Models 1, 2 and 3

Figure 9 illustrates the variation of the maximum flow velocity of liquid metal in the molten pool region with laser heating and cooling duration during the polishing process. With different initial surface characteristics, the polished surface starts to partially melt at a moment of about 0.3 ms. Thereafter, at 0.3~0.5 ms, the molten pool velocity gradually increases with heating time. It can be noticed that Model 1 has the smallest velocity (2.11 m/s) compared to Model 2 and Model 3 at 0.5 ms. The maximum velocity in the melting region (3.13 m/s) was then reached at 1 ms. Model 2 has the second highest velocity (2.76 m/s) at 0.5 ms compared to Models 1 and 3, followed by a maximum melt velocity (3.15 m/s) at 3.5 ms. Model 3 has the maximum velocity (3.13 m/s) and the maximum melt flow velocity (3.15 m/s) at 1 ms. In the cooling phase, the high cooling rate of the molten pool temperature is accompanied by a sharp increase in the melt dynamic viscosity, which results in a cliff-like decrease in the melt flow rate. Models 1 and 3 have a high dynamic viscosity close to 0 at 3.59 ms. They have a molten pool survival lifetime of about 0.09 ms. Model 2 has a flow rate close to 0 at 3.62 ms and a molten pool survival lifetime of about 0.12 ms. The results show that the overall molten pool velocity under different initial profiles is inverted "U" profile and quasi-stable when the melt reaches a certain velocity. In addition, different surface features influence the moment and location of the maximum velocity of the molten pool. It also has an effect on the survival life of the molten pool, which further affects the final polished morphology.

The variation of the molten pool depth and width in the polished area with the laser radiation duration for different initial morphological features, as shown in Figure 10. It can be found that the molten pool depths of Model 2 at radiation times of 0.5~1.4 ms are slightly greater than those of Models 1 and 3. At 2~3 ms, the melting depths of Models 1 and 3 showed a decreasing trend. In contrast, the Model 2 melt depth increases with time but more slowly. Finally, Models 1 and 3 start to rise at 3.5 ms and the melting depth increases to 58 μm and 47 μm, respectively. But the Model 2 melting depth slightly decreased to 49 μm. The results show an overall increasing trend of molten pool depth with time during the polishing process. However, the different surface morphological features affect the mass flow of the melt in the z-direction, thus, affecting the heat transfer inside the molten pool. It further causes the melt depth to fluctuate with the laser radiation time. Additionally, the general pattern of the molten pool width parallel to the polishing direction with time expressed in Figure 10b is similar to that of the melt depth. Furthermore, as the cooling rate of the molten pool tail temperature is greater than the rate of expansion displacement of the melt front. It can be found that the melt widths of Models 1, 2 and 3 all drop locally during the polishing process.

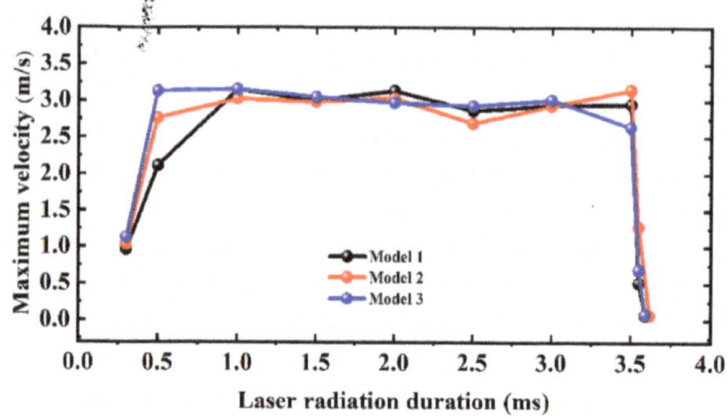

Figure 9. Variation of the maximum velocity of liquid metal in the polished area with radiation duration.

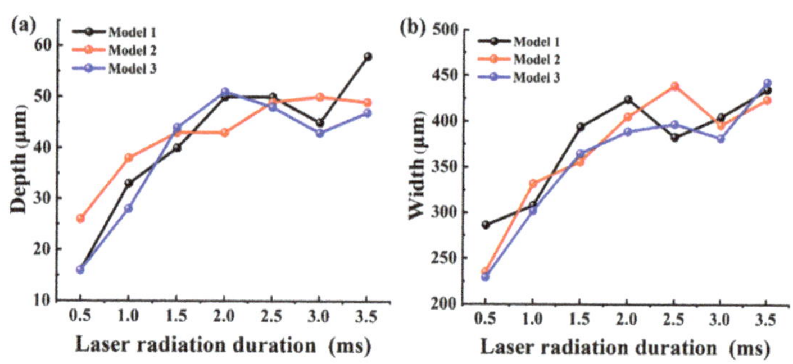

Figure 10. Variation of molten pool depth and width in the polished area with radiation duration: (**a**) melting depth; (**b**) melting width.

Figure 11 demonstrates the profile height of the polished surface in different initial morphologies. It can be seen from the figure that the height of the half-wave profile at the starting position of the surface after LP is the largest (24 μm) for Model 1, the second (13.5 μm) for Model 3 and the smallest (1.6 μm) for Model 2. It maps the melt to the strongest Marangoni convection at this location. In addition, it can be found that the surface curvature at the starting position of the unpolished morphology of Model 2 is the largest (see Figure 1). It illustrates the normal smoothing effect of the capillary force is the most obvious, which basically eliminates the surface bulge. It can be further observed that the polished surface is not eliminated by the normal and tangential thermocapillary forces despite the reduction in the number of peaks and the height of the profile. At the same time, the local surface height is higher than the initial surface, which is due to the short time that the molten pool is in the molten state and the capillary force is too late to normalize the large curvature surface. Meanwhile, the Marangoni convection at the tail of the molten pool causes the melt to flow along the center of the beam towards the edge. As a result, the peaks are not effectively smoothed, and the surface formed local bulge after polishing. In addition, the different surface curvature due to the difference in initial morphology affects the dominant role of capillary forces on the molten pool flow as well as the dominant role of thermocapillary forces after the molten pool is fully developed.

Finally, it can be noted that due to the Marangoni convection of the double spiral nest, the polished end position forms a depression larger than the diameter of the beam. The depression width and depth of Model 2 are the smallest, about 312 μm and 12 μm. Model 3 is the second, about 380 μm and 13 μm. In addition, Model 1 is the largest, about 520 μm and 26 μm. It can also be seen from the above data that the velocity and direction of the flow to the valleys during peaks melting is relatively stable due to the relatively uniform distribution of the surface morphology at the initial surface end of Model 2. The fluctuations generated on the surface of the molten pool are smaller, and a better effect of melt peak filling can be achieved. Thus, the formed depressions are uniform in profile and small in extent. In contrast, the peaks distribution at the end position of the initial surface of Model 1 has an overall negative curvature distribution. This is coupled with the effect of Marangoni convection tangential along the surface. It results in a large depression depth and width. The above results indicate that the initial surface features with the uniform distribution can enhance the smoothing effect.

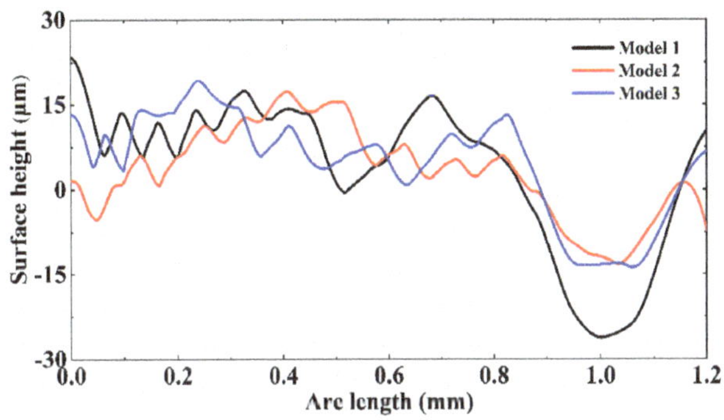

Figure 11. Profile height of the polished surface.

4.8. Experimental Validation

The surface morphology of the single-line polished tracks and comparison between simulated surface height profile and experimental laser polished profiles for Models 1, 2 and 3, as shown in Figure 12. Fourier filter is performed on the 3D morphology of the polished surface so that ensuring the accuracy of the experimental surface profiles. The processing method is the same as described in Section 2.2. The red lines in the wire frame are the single-line polished tracks morphology for Models 1, 2 and 3. The yellow lines indicate the surface line profiles along the center of polished track area (see Figure 12a,c,e). Correspondingly, the red lines indicate the polished surface profiles predicted by numerical simulation, the blue lines represent polished surface profiles by experiments as well as the black represent the initial surface morphology. From the overall profiles of the polished surfaces, the surface asperities were not completely eliminated by means of CW fiber laser polishing. Simultaneously, the positive and the negative curvature profiles were formed nearby the starting position and at the edges of the polished surface. Additionally, it can be observed that the experimental depth of the depressions for Models 1, 2 and 3 are about 24 μm, 14 μm and 15 μm at 1 mm of the polished surface. Correspondingly, the numerical prediction with the depth of about 26 μm, 12 μm and 13 μm. The results indicate that the models surface profile agree reasonably well with the experimental data.

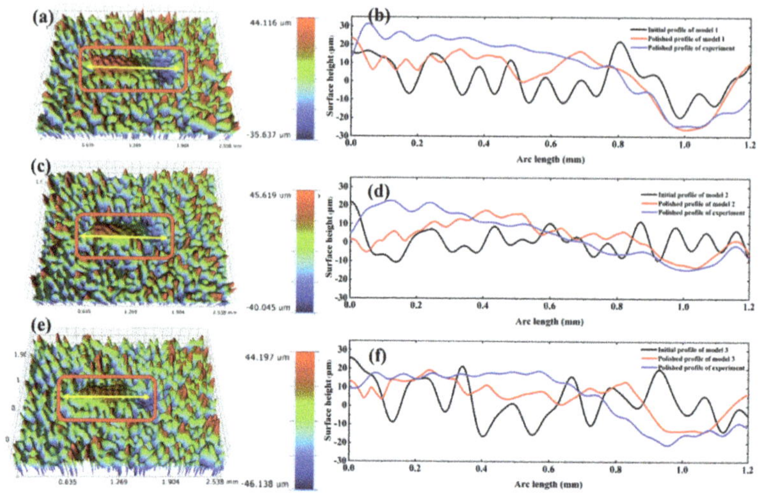

Figure 12. The surface morphology of the single-line polished tracks and comparison between simulated surface height profile and experimental laser polished profiles for Models 1, 2 and 3. (**a**) Polished optical morphology of Model 1; (**b**) Surface profile height of Model 1; (**c**) Polished optical morphology of Model 2; (**d**) Surface profile height of Model 2; (**e**) Polished optical morphology of Model 3; (**f**) Surface profile height of Model 3.

5. Conclusions

In this work, we developed a transient 2D model coupled of heat transfer and fluid fields based on moving heat source. It is illustrated that the effect of different initial morphologies on the evolution of the melt hydrodynamics during LP. The following conclusions can be made:

(1) The model demonstrated that the complex evolution of the melt hydrodynamics involving heat conduction, thermal convection, thermal radiation, melting and solidification during laser polishing.
(2) The uniformity of the initial surface peaks and valleys distribution is positively correlated with the smoothing quality of the polished surface, but has less effect on temperature field, velocity field, as well as melt depth and width of the molten pool.
(3) The surface rough profiles are not completely eliminated by capillary and thermocapillary forces due to the high cooling rate of the molten pool, resulting in the formation of secondary surface roughness. It was revealed that the short lifetime of the molten pool is the main reason why the surface bumps are not completely eliminated.
(4) The numerical prediction of the depressions for Models 1, 2 and 3 are approximate 26 μm, 12 μm and 13 μm at about 1 mm on the polished surface. Accordingly, the experimental molten pool depths are about 24 μm, 14 μm and 15 μm as well as the errors are approximately 8.3%, 14.3% and 13.3%, respectively.
(5) The model not only predicts the morphological evolution of different surfaces from rough to smooth in laser polishing, but also can be suitable for guiding the optimization of polishing parameters such as laser power and scanning speed. Additionally, this model can be applied to most metallic materials in laser polishing.

Author Contributions: Manuscript writing, research design, data analysis: K.L.; Funding acquisition, project administration: Z.Z. and H.Z. (Houming Zhou); Article analysis, review and revision: H.Z. (Hao Zhou) and J.Y.; Making charts, literature search: W.Z. and G.Z. All authors have read and agreed to the published version of the manuscript.

Funding: This research was funded by Natural Science Foundation of Hunan Province (2020JJ4580), Shenzhen Science and Technology Plan Project: Rapid Laser Polishing Mechanism and Process Research of Surface (No.JCYJ20170817112445033), Guangdong Province Higher Vocational Colleges and Schools Pearl River Scholar Funded Scheme 2018, Innovation Team Project on University of Guangdong Province (No.2018GKCXTD001), Technology Innovation Platform Project of Shenzhen Institute of Information Technology 2020 (PT2020E002), Hunan innovative province construction special major landmark innovation demonstration project, Changsha Zhuzhou Xiangtan landmark engineering technology project (2019XK2303), Xiangtan science and technology project (ZD-ZD20191007), Natural Science Foundation of China (51775470), Shenzhen Science and Technology Plan Project (GJHZ20180416164715805), Department of Science and Technology of Guangdong Province (2020B1212060067).

Data Availability Statement: Not applicable.

Conflicts of Interest: The authors declare that they have no known competing financial interest or personal relationship that could have appeared to influence the work reported in this paper.

Nomenclature

LP	Laser polishing
PAM	Plasma assisted milling
FEM	Finite element method
ALE	Arbitrary Lagrangian-Eulerian
SSM	Surface shallow melting
2D	Two dimensional
f_{cr}	Critical frequency (Hz)
ρ	Density (kg m^{-3})
t	Laser radiation duration (ms)
T	Temperature (K)
$\vec{u}$	Velocity (m s^{-1})
K	Thermal conductivity (Wm^{-1} K^{-1})
$\vec{u}_m$	Mesh velocity
u_{mat}	Material velocity (m s^{-1})
p	Pressure (Pa)
I	Identity matrix
M	Dynamic viscosity (Pa s)
F_V	Body force (N m^{-3})
T_{ref}	Reference temperature (K)
ρ_{ref}	Reference density (kg m^{-3})
β	Thermal expansion coefficient (K^{-1})
g	Gravity constant (N kg^{-1})
C_p	Specific heat (Jkg^{-1} K^{-1})
f_L	Liquid fraction
T_s	Solidus temperature (K)
T_l	Liquidus temperature (K)
T_m	Melting temperature (K)
T_b	Boiling temperature (K)
T_a	Ambient temperature (K)
ρ_s	Solidus density (kg m^{-3})
ρ_l	Liquidus density (kg m^{-3})
k_s	Solidus thermal conductivity (Wm^{-1} K^{-1})
k_l	Liquidus thermal conductivity (Wm^{-1} K^{-1})
$C_{p\text{-}s}$	Solidus specific heat (Jkg^{-1} K^{-1})
$C_{p\text{-}l}$	Liquidus specific heat (Jkg^{-1} K^{-1})
H	Convective coefficient (Wm^{-2} K^{-1})
$\partial\gamma/\partial T$	Temperature derivative of surface tension (Nm^{-1} K^{-1})
L_m	Latent heat of melting (Jkg^{-1})

E	Emissivity
α_0	Absorptivity
σ	Stefan-Boltzmann constant
u_r	Fluid flow velocity along the r direction
u_z	Fluid flow velocity along the z direction
γ	Surface tension coefficient (N m^{-1})
κ	Surface curvature (m^{-1})
$\vec{n}$	Normal vector
$\vec{t}$	Tangential vector
P	Laser power (W)
r_0	Laser beam radius (m)
M^2	Laser beam quality
R	The r-component in the cylindrical coordinate system
V	Laser moving velocity (mm s^{-1})
$f(r^*)$	Laser radiation area
Q_s	Stationary laser energy density (J cm^{-2})
Q_m	Moving laser energy density (J cm^{-2})
t_h	Laser heating duration (ms)
t_c	Cooling duration (ms)

References

1. Singh, P.; Pungotra, H.; Kalsi, N.S. On the characteristics of titanium alloys for the aircraft applications. *Mater. Today Proc.* **2017**, *4*, 8971–8982. [CrossRef]
2. Gurrappa, I. Characterization of titanium alloy Ti6Al4V for chemical, marine and industrial applications. *Mater. Charact.* **2003**, *51*, 131–139. [CrossRef]
3. Giannatsis, J.; Dedoussis, V. Additive fabrication technologies applied to medicine and health care: A review. *Int. J. Adv. Manuf. Technol.* **2009**, *40*, 116–127. [CrossRef]
4. Pimenov, D.Y.; Guzeev, V.I. Mathematical model of plowing forces to account for flank wear using FME modeling for orthogonal cutting scheme. *Int. J. Adv. Manuf. Technol.* **2017**, *89*, 3149–3159. [CrossRef]
5. Pimenov, D.Y.; Mia, M.; Gupta, M.K.; Machado, A.R.; Tomaz, Í.V.; Sarikaya, M.; Wojciechowski, S.; Mikolajczyk, T.; Kapłonek, W. Improvement of machinability of Ti and its alloys using cooling-lubrication techniques: A review and future prospect. *J. Mater. Res. Technol.* **2021**, *11*, 719–753. [CrossRef]
6. Liu, S.Y.; Yung, C.S. Additive manufacturing of Ti6Al4V alloy: A review. *Mater. Des.* **2019**, *164*, 107752. [CrossRef]
7. Ukar, E.; Lamikiz, A.; Martínez, S.; Estalayo, F.; Tabernero, I. Laser Polishing of GGG70L Cast Iron with 2D scan-head. *Pro. Eng.* **2013**, *63*, 53–59. [CrossRef]
8. Martínez, S.; Lamikiz, A.; Ukar, E.; Calleja, A.; Arrizubieta, J.A.; López de Lacalle, L.N. Analysis of the regimes in the scanner-based laser hardening process. *Opt. Laser. Eng.* **2017**, *90*, 72–80. [CrossRef]
9. Zhou, J.; Liao, C.H.; Shen, H.; Ding, X.H. Surface and property characterization of laser polished Ti6Al4V. *Surf. Coat. Technol.* **2019**, *380*, 125016. [CrossRef]
10. Xu, Z.; Ouyang, W.T.; Liu, Y.F.; Jiao, J.K.; Liu, Y.Z.; Zhang, W.W. Effects of laser polishing on surface morphology and mechanical properties of additive manufactured TiAl components. *J. Manuf. Process.* **2021**, *65*, 51–59. [CrossRef]
11. López de Lacalle, L.N.; Sa'nchez, J.A.; Lamikiz, A.; Celaya, A. Plasma assisted milling of heat-resistant superalloys. *J. Manuf. Sci. Eng.* **2004**, *126*, 274–285. [CrossRef]
12. López de Lacalle, L.N.; Lamikiz, A.; Celaya, A. Simulation of plasma assisted milling of heat resistant alloys. *Int. J. Simul. Model.* **2002**, *1*, 5–15.
13. Mai, T.A.; Lim, G.C. Micromelting and its effects on surface topography and properties in laser polishing of stainless steel. *J. Laser. App.* **2004**, *16*, 221–228. [CrossRef]
14. Perry, T.L.; Werschmoeller, D.; Duffie, N.A.; Li, X.C.; Pfefferkorn, F.E. Examination of selective pulsed laser micropolishing on microfabricated nickel samples using spatial frequency analysis. *J. Manuf. Sci. Eng.* **2009**, *131*, 021002. [CrossRef]
15. Perry, T.L.; Werschmoeller, D.; Li, X.C.; Pfefferkorn, F.E.; Duffie, N.A. The effect of laser pulse duration and feed rate on pulsed laser polishing of micro fabricated nickel samples. *J. Manuf. Sci. Eng.* **2009**, *131*, 031002. [CrossRef]
16. Perry, T.L.; Werschmoeller, D.; Li, X.C.; Pfefferkorn, F.E.; Duffie, N.A. Pulsed laser polishing of micro-milled Ti6Al4V samples. *J. Manuf. Process.* **2009**, *11*, 74–81. [CrossRef]
17. Vadali, M.; Ma, C.; Duffie, N.A.; Li, X.; Pfefferkorn, F.E. Pulsed laser micro polishing: Surface prediction model. *J. Manuf. Process.* **2012**, *14*, 307–315. [CrossRef]
18. Ukar, E.; Lamikiz, A.; Martínez, S.; Tabernero, I.; López de Lacalle, L.N. Roughness prediction on laser polished surfaces. *J. Mater. Process. Technol.* **2012**, *212*, 1305–1313. [CrossRef]

19. Wang, Q.H.; Morrow, J.D.; Ma, C.; Duffie, N.A.; Pfefferkorn, F.E. Surface prediction model for thermocapillary regime pulsed laser micro polishing of metals. *J. Manuf. Process.* **2015**, *20*, 340–348. [CrossRef]
20. Sim, B.; Kim, W. Melting and Dynamic-surface Deformation in Laser Surface Heating. *Int. J. Heat Mass Transfer.* **2005**, *48*, 1137–1144. [CrossRef]
21. Ma, C.; Vadali, M.; Duffie, N.A.; Pfefferkorn, F.E.; Li, X. Melt pool flow and surface evolution during pulsed laser micro polishing of Ti6Al4V. *J. Manuf. Sci. Eng.* **2013**, *135*, V001T01A054. [CrossRef]
22. Zhang, C.; Zhou, J.; Shen, H. Role of Capillary and Thermocapillary Forces in Laser Polishing of Metals. *J. Manuf. Sci. Eng.* **2017**, *139*, 041019. [CrossRef]
23. Shen, H.; Pan, Y.; Zhou, J.; Yao, Z. Forming mechanism of bump shape in pulsed laser melting of stainless steel. *J. Heat Transfer.* **2017**, *139*, 062301. [CrossRef]
24. Li, K.; Zhao, Z.Y.; Zhou, H.M.; Zhou, H.; Jin, J.C. Numerical analyses of molten pool evolution in laser polishing Ti6Al4V. *J. Manuf. Process.* **2020**, *58*, 574–584. [CrossRef]
25. Li, K.; Zhou, H.M.; Zhao, Z.Y.; Zhou, H.; Yin, J. A study on transient molten pool dynamics in laser polishing of Ti6Al4V using numerical simulation. *J. Manuf. Process.* **2021**, *65*, 478–490. [CrossRef]
26. Sharma, S.; Mandal, V.; Ramakrishna, S.A.; Ramkumar, J. Numerical simulation of melt hydrodynamics induced hole blockage in Quasi-CW fiber laser micro-drilling of TiAl6V4. *J. Mater. Process. Technol.* **2018**, *262*, 131–148. [CrossRef]
27. Mills, K.C. *Recommended Values of Thermophysical Properties for Selected Commercial*; Woodhead Publishing Ltd: Cambridge, UK, 2002.
28. Ramos, J.A.; Murphy, J.; Wood, K.; Bourell, D.L.; Beaman, J.J. Surface roughness enhancement of indirect-SLS metal parts by laser surface polishing. In Proceedings of the 12th Solid Freeform Fabrication Symposium, Austin, TX, USA, 6–8 August 2001; pp. 28–38.
29. Ramos, J.A.; Bourell, D.L.; Beaman, J.J. Surface over-melt during laser polishing of indirect-SLS metal parts. *MRS Proc.* **2002**, *758*, 53–61. [CrossRef]
30. Ramos, J.A.; Bourell, D.L.; Beaman, J.J. Surface characterization of laser polished indirect-SLS parts. In Proceedings of the Solid Freeform Fabrication Symposium, Austin, TX, USA, 5–7 August 2002; pp. 554–562. [CrossRef]
31. Ramos, J.A.; Bourell, D.L. Reducing surface roughness of metallic freeform-fabricated parts using non-tactile finishing methods. *Int. J. Mater. Prod. Technol.* **2004**, *21*, 297–316. [CrossRef]
32. Zhang, W.K.; Dai, W.; Zheng, Z.Z.; Li, J.J.; Deng, T.T. Numerical simulation and verification of free surface evolution of laser polished H13 die steel. *Chin. J. Lasers* **2019**, *46*, 0702002. [CrossRef]
33. Morville, S.; Carin, M.; Muller, M.; Gharbi, M.; Peyre, P.; Carron, D.; Lemasson, L.F.; Fabbro, R. 2D axial-symmetric model for fluid flow and heat transfer in the melting and resolidification of a vertical cylinder. In Proceedings of the 2010 COMSOL Conference, Paris, France, 17–19 November 2010.
34. Fontes, E. Two Methods for Modeling Free Surfaces in COMSOL Multiphysics. 2018. Available online: https://www.comsol.com/blogs/two-methods-for-modeling-free-surfaces-in-comsol-multiphys-ics/ (accessed on 15 May 2018).
35. Modeling Free Surfaces in COMSOL Multiphysics with Moving Mesh. 2018. Available online: https://cn.comsol.com/blogs/modeling-free-surfaces-in-comsol-multiphysics-with-moving-mesh/ (accessed on 31 May 2018).
36. Zhou, Y.Q.; Zhao, Z.Y.; Zhang, W.; Xiao, H.B.; Xu, X.M. Experiment study of rapid laser polishing of freeform steel surface by dual-beam. *Coatings* **2019**, *9*, 324. [CrossRef]
37. MFSC 700W-1500W Series Continuous Fiber Laser user Manual Reference. 2021. Available online: http://www.maxphotonics.com/Cnp/35.html/ (accessed on 6 May 2021).
38. Courtois, M.; Carin, M.; Masson, P.L.; Gaied, S.; Balabane, M. A new approach to compute multi-reflections of laser beam in a keyhole for heat transfer and fluid flow modelling in laser welding. *J. Phys. D Appl. Phys.* **2013**, *46*, 505305. [CrossRef]
39. Zhang, Y.M.; Shen, Z.H.; Ni, X.W. Modeling and simulation on long pulse laser drilling processing. *Int. J. Heat Mass Tran.* **2014**, *73*, 429–437. [CrossRef]
40. Shen, H.; Feng, D.; Yao, Z. Modeling of underwater laser drilling of alumina. *J. Manuf. Sci. Eng.* **2016**, *139*, 041008. [CrossRef]

micromachines

Article

Study on Manufacturing Technology of Ag-8.5Au-3.5Pd Fine Alloy Wire

Jun Cao [1,*], Junchao Zhang [1], Baoan Wu [2], Huiyi Tang [2], Changchun Lv [3], Kexing Song [4], Guannan Yang [5], Chengqiang Cui [5] and Yangguang Gao [3]

1 School of Mechanical and Power Engineering, Henan Polytechnic University, Jiaozuo 454000, China; zjc_xlyx@sina.com
2 Chongqing Materials Research Institute Co., Ltd., Chongqing 400700, China; wubaoan@163.com (B.W.); hytang320@163.com (H.T.)
3 Henan Youk Electronic Material Co., Ltd., Jiyuan 454650, China; ykdz0391@163.com (C.L.); caolinc@163.com (Y.G.)
4 School of Materials Science and Engineering, Henan University of Science and Technology, Luoyang 471000, China; kxsong@haust.edu.cn
5 State Key Laboratory of Precision Electronic Manufacturing Technology and Equipment, Guangdong University of Technology, Guangzhou 510006, China; ygn@gdut.edu.cn (G.Y.); cqcui@gdut.edu.cn (C.C.)
* Correspondence: cavan@hpu.edu.cn

Abstract: The performance of Ag-8.5Au-3.5Pd alloy wire after cold deformation and annealing were analyzed by SEM (scanning electron microscope), strength tester and resistivity tester. The processing process and performance change characteristics of Ag-8.5Au-3.5Pd alloy wire were studied. The results show that alloy wire grains gradually form a fibrous structure along with the increase in deformation. The strength of the wire increases with the increase in deformation rate, but the increase trend becomes flat once the deformation rate is higher than 92.78%; the resistivity of Ag-8.5Au-3.5Pd alloy wire decreases with the increase in annealing temperature, reaching minimum (2.395×10^{-8} Ω·m) when the annealing temperature is 500 °C; the strength of Ag-8.5Au-3.5Pd alloy wire decreases with the increase in annealing temperature. When the annealing temperature is 500 °C, the strength and elongation of the φ0.2070 mm Ag-8.5Au-3.5Pd alloy wire are 287 MPa and 25.7%, respectively; the fracture force and elongation of φ0.020 mm Ag-8.5Au-3.5Pd alloy wire are 0.0876 N and 14.8%, respectively. When the annealing temperature is 550 °C, the metal grains begin to grow and the mechanical performance decrease; the φ0.020 mm Ag-8.5Au-3.5Pd alloy wire have good surface quality when the tension range is 2.5–3.0 g.

Keywords: alloy wire; processing rate; annealing; strength; elongation; resistivity; tension

Citation: Cao, J.; Zhang, J.; Wu, B.; Tang, H.; Lv, C.; Song, K.; Yang, G.; Cui, C.; Gao, Y. Study on Manufacturing Technology of Ag-8.5Au-3.5Pd Fine Alloy Wire. *Micromachines* **2021**, *12*, 938. https://doi.org/10.3390/mi12080938

Academic Editor: Nikolaos Tapoglou

Received: 31 May 2021
Accepted: 5 August 2021
Published: 9 August 2021

Publisher's Note: MDPI stays neutral with regard to jurisdictional claims in published maps and institutional affiliations.

1. Introduction

With the development of semiconductor devices and integrated circuits toward multilead packaging, high integration and miniaturization, more attention has been paid to ultra-fine, low-cost, high-temperature-resistant wires [1–3]. Bonding gold wires are affected by the shortcomings of high-cost, low strength and brittle intermetallics and are gradually to be replaced by other materials [4,5]. The cost of copper wire is low, but it still has some drawbacks such as easy oxidation and high hardness [6–8]. Aluminum wires have weaker tensile strength and heat resistance than gold wires, so they are prone to collapse and droop [9]. Nowadays, there are many kinds of bonding materials, among which the silver base alloy bonding wires exhibit excellent mechanical properties, good oxidation resistance and high reliability, and are cost effective. As a result, they can limit light attenuation and improve the conversion rate in light-emitting diode (LED) packaging. Because of these advantages, these wires are employed widely in integrated circuit and LED packaging [10–17]. However, for pure Ag wire, due to high thermal conductivity and low strength of high temperature, the parameter range is small during the bonding process, and has high failure probability of ball bonding points under high temperature conditions, thus reducing production efficiency and service life of high-power LED devices [18–23]. High-performance silver-based alloy wires obtained by alloying is an effective method to improve the performance of silver wire. Au (gold) and Pd (palladium) elements have similar performance to Ag and can be infinitely soluble with each other. The addition of Au and Pd elements can improve the strength and high temperature stability of silver wire [24], and increase the parameter window range and interface bonding strength during bonding process [25], and inhibit the growth of intermetallics at interface (especially under high temperature and humidity conditions), which is conducive to further increasing interface reliability [26] and enhancing device life. The research and development of silver-based alloy wire with Au and Pd solves many problems in the application of single-element wire such as gold wire, copper wire and silver wire, and have broad application prospects in chip packaging such as high-density, large-scale integrated circuits and high-power LED. In recent years, many scholars have carried out a lot of research on Ag/Au/Pd alloy wires. Feng, D. et al. [27] studied the thermodynamic performance of Ag-Au-Pd alloy by electrochemical methods and concluded that the thermodynamic stability of Ag-Au-Pd is significantly higher than Ag. Guo, R. et al. [28] studied the intermetallics at the interface of Ag-8Au-3Pd alloy wire and concluded that $AuAl_2^+$ (Au, Ag) 4Al and Ag_2Al intermetallics were generated at the interface of Ag-8Au-3Pd/Al, and the Ag_2Al intermetallics layer could effectively prevent the diffusion of Au atoms to the interface of Al. Cao, J. et al. [29,30] studied the cold deformation and annealing process of Ag-4Pd alloy wire and summarized the influence of the performance and structure of Ag-4Pd alloy on strength. It is concluded that twin crystal structure appears during the annealing of the Ag-4Pd alloy wire, and twin nucleation and subcrystalline annexation to large nucleation is the main nucleation mode, and the length of the heat-affected zone is shorter than Ag wire. Most of them involve the research on the reliability of Ag/Au/Pd alloy wire and the Ag/Al interface, while there is little discussion on wire performance and processing technology of Ag/Au/Pd alloys. In this paper, the influence of the deformation rate and annealing on the performance of the Ag-8.5Au-3.5Pd alloy wire during processing is studied, and the processing technology of the Ag-8.5Au-3.5Pd alloy wire is further explored, which provides a theoretical basis for the manufacture of Ag-8.5Au-3.5Pd alloy wire.

2. Test Materials and Methods

2.1. Test Materials

φ8 mm diameter Ag-8.5Au-3.5Pd alloy rod and drawing dies with a diameter range of 8.00–0.020 mm.

2.2. Test Method

The φ8 mm Ag-8.5Au-3.5Pd melting cast alloy rod was cold deformed by large drawing, medium drawing, fine drawing and ultra-fine drawing. The structure, outline size and hole size of drawing dies are shown in Figure 1, Tables 1 and 2, respectively. Firstly, the alloy wire is machined to φ1.1008 mm on the single-die drawing machine, and the compression rate and drawing speed of the alloy wire are 10% and 20%, 10 m/min and 30 m/min in the range of dies with a hole diameter of 8.00–5.2488 mm and 4.6947–1.1008 mm, respectively. The drawing solution is 396V2 water-soluble solution and the concentration is 20% (the composition of the drawing solution is 50% polyethylene glycol, 30% dehydrated sorbitol monooleate polyoxyethylene ether and 20% water). A 10-times magnifying glass is used to inspect the surface of the wire at any time during rough drawing, and the surface is required to be smooth. Then, the φ1.1008 mm alloy wire is machined to φ0.2070 mm on the LH160 Mid-Drawing wire drawing machine, and the compression rate and drawing speed of the alloy wire is 13% and 200 m/min. In this process, the drawing solution is water-soluble, and the concentration is 5% and the temperature is 40 °C. The φ0.2070 mm alloy wire is heat treated according to the temperature and time shown in Table 3, and the fluctuation range of the annealing temperature is ±2 °C. The annealing tube is made of quartz glass and the length is 2000 mm. The tension of the wire is controlled by angular displacement sensor in the annealing equipment and the tension range is 1.0–10.0 g. High purity N_2 protection is used in the annealing process. In the process of rough drawing, both the alloy wires of different diameters (φ8.0 mm, φ5.2488 mm, φ3.3592 mm, φ2.1499 mm, φ1.1008 mm, φ0.2070 mm) after cold deformation and the φ0.2070 mm alloy wire after annealing are sampled. The alloy samples were corroded for 3~5 s by the solution (30% mass fraction hydrogen peroxide and 20% mass fraction ammonia mixed in the same proportion). The solution should be used immediately once prepared. The microstructure morphology and grain size of the alloy wire under different deformation variables were observed by JEOL JSM-6700F SEM and ZEISS SEM. Finally, mechanical and electrical performance of Ag-8.5Au-3.5Pd alloy wires that annealed at different temperatures were tested on the KDDII-0.01 tension machine and ZX01 double-arm electric bridge, respectively, and the effects of different deformation amounts and annealing temperatures on the wire performance were studied. The tensile test samples are 100 mm in length and 10 mm/min in tension speed. The electrical performance test sample length is 1000 mm; the resistance value is measured by 4 probe method with the ZDCY-80 intelligent resistance tester and then according to the formula $\rho = R \times S/L$ calculates the resistivity of the alloy wire (ρ is resistivity, L is wire length, S is section area).

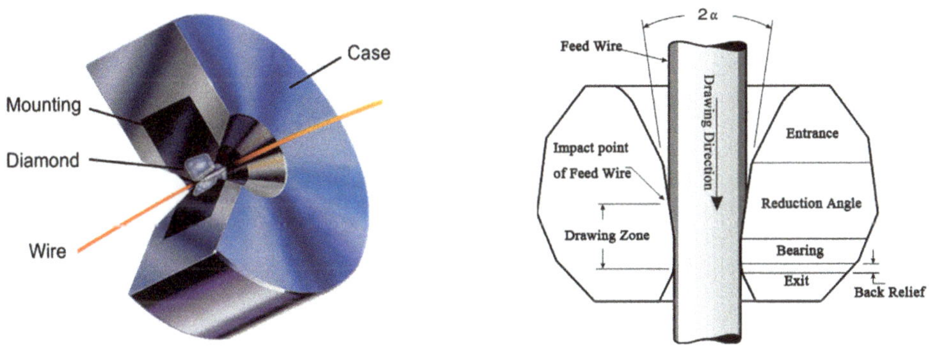

Figure 1. Drawing die structure.

Table 1. Wire drawing die dimensions.

Step	Range of Dies Hole Size (mm)	Dies Sleeve Size (mm)		
		External Diameter	Thickness	Tolerance
Rough drawing	5.210~7.628	42	27	+0, −0.1
	1.513~4.798	42	27	+0, −0.1
Medium drawing	0.5480~1.4138	25	8	+0, −0.1
	0.1985~0.5121	25	8	+0, −0.1
Fine drawing 1	0.11380–0.19170	25	6	+0, −0.02
Fine drawing 2	0.06525–0.010991	25	6	+0, −0.02
Fine drawing 3	0.03874–0.06032	25	6	+0, −0.02
Ultra-fine drawing 1	0.03035~0.03780	25	6	+0, −0.02
Ultra-fine drawing 2	bellow 0.03	25	6	+0, −0.02
Ultra-fine drawing 3	bellow 0.03	25	6	+0, −0.02

Table 2. Wire drawing die dimensions.

Step	Range of Dies Hole Size (mm)	Aperture Tolerance (mm)	Roundness (mm)	Entry Angle (°)	Bearing Length (%D)	Exit Zone Angle (°)
Rough drawing	5.210~7.628	+0, −0.004	0.004	15° ± 2°	30 ± 10	15° ± 5°
	1.513~4.798	+0, −0.003	0.003	15° ± 2°	30 ± 10	15° ± 5°
Medium drawing	0.5480~1.4138	+0, −0.002	0.001	15° ± 2°	20~30	15° ± 5°
	0.1985~0.5121	+0, −0.001	0.001	15° ± 2°	20~40	15° ± 5°
Fine drawing 1	0.11380~0.19170	+0, −0.0005	0.0005	13° ± 2°	50 ± 10	15° ± 5°
Fine drawing 2	0.06525~0.010991	+0, −0.0003	0.0003	13° ± 2°	50 ± 10	15° ± 5°
Fine drawing 3	0.03874~0.06032	+0, −0.0002	0.0002	13° ± 2°	50 ± 10	15° ± 5°
Ultra-fine drawing 1	0.03035~0.03780	+0, −0.0002	0.0002	13° ± 2°	50 ± 10	15° ± 5°
Ultra-fine drawing 2	bellow 0.03	+0, −0.0002	0.0002	13° ± 2°	50 ± 10	15° ± 5°
Ultra-fine drawing 3	bellow 0.03	+0, −0.0002	0.0002	13° ± 2°	50 ± 10	15° ± 5°

Table 3. Annealing process parameters of Ag-8.5Au-3.5Pd alloy wire.

Number	Temperature (°C)	Time (s)	Remarks
1	200	2.4	The annealing tube length is 2.0 m and the annealing speed is 50 m/min.
2	300	2.4	
3	400	2.4	
4	450	2.4	
5	500	2.4	
6	550	2.4	

The φ0.2070 mm Ag-8.5Au-3.5Pd alloy wire treated by optimized annealing conditions was cold-worked on the SPS-100 drawing machine. The drawing process includes two passes, and the diameter ranges of the dies are φ0.2070–φ0.1158 mm and φ0.11169–φ0.0648 mm in the first and second pass, respectively; the compression rate is 7%, and the drawing speed is 500–600 m/min. The drawing solution is water-soluble, which concentration is 2.0% and the temperature is 40 °C. Then, the φ0.06481 mm alloy wire is drawn by the MK-30 no-sliding wire drawing machine. The drawing process also includes two passes. In the first pass, the rang of the die diameter is φ0.06250–φ0.03761 mm and the compression rate is 7%. In the second pass, the range of the die diameter is φ0.03636–φ0.02513 mm and the compression rate is 6.5%. The drawing speed is 500–600 m/min. The drawing solution is 396V2 water-soluble solution which concentration is 0.5% and the temperature is 40 °C. Finally, the φ0.02513 mm alloy wire is processed to φ0.01962 mm on the WSS-11 Micro wire drawing machine; the compression rate is 6.0% and the drawing speed is 300–400 m/min. The drawing solution concentration is 0.5% and the temperature is 40 °C.

3. Results

3.1. Analysis of the Influence of Cold Deformation on the Microstructure of Ag-8.5Au-3.5Pd Alloy Wire

Figure 2 shows the microstructures in different deformation (φ8.0 mm, φ5.2488 mm, φ3.3592 mm, φ2.1499 mm, φ1.1008 mm and φ0.2070 mm) of the Ag-8.5Au-3.5Pd alloy wire after multi-pass drawing. It can be seen that during plastic deformation, the grains are gradually straightened and refined, and finally the microstructures present fibrous (as shown in the microstructure of φ0.02070 mm size) with the increase in deformation of the wire. In addition, the number of grains and the degree of microstructure distortion also increase.

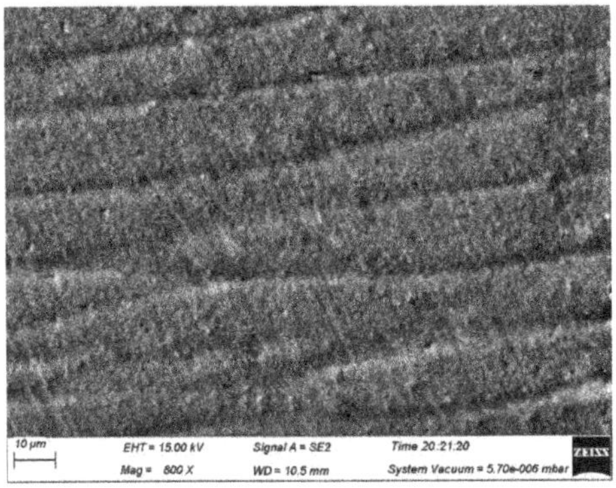

(**a**) Initial coarse grains in longitudinal section. (φ8.0 mm)

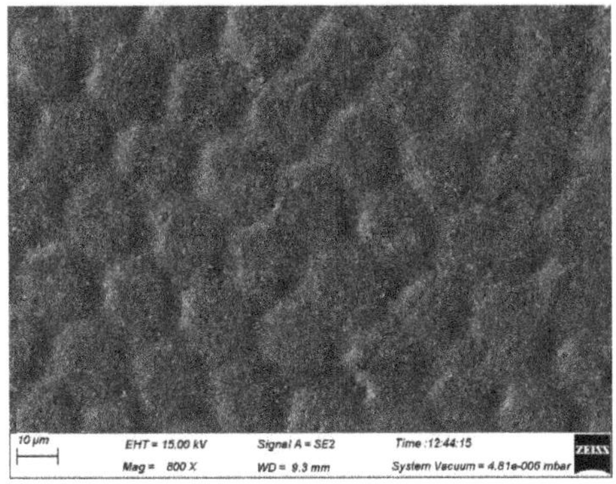

(**b**) Large diameter grains in cross section. (φ8.0 mm)

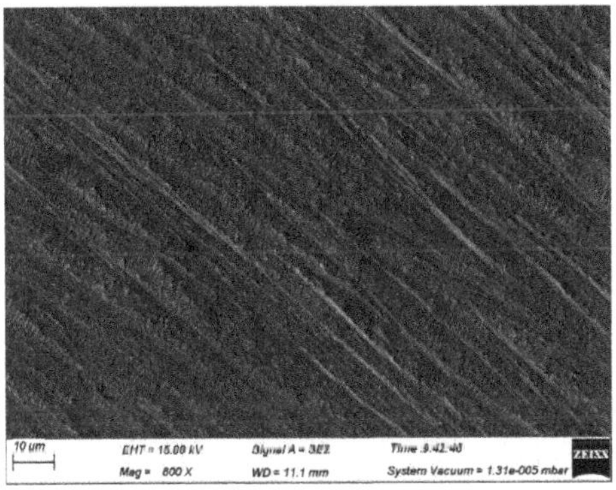

(**c**) The coarse and large grains are refined and lengthened in longitudinal section. (φ3.3592 mm)

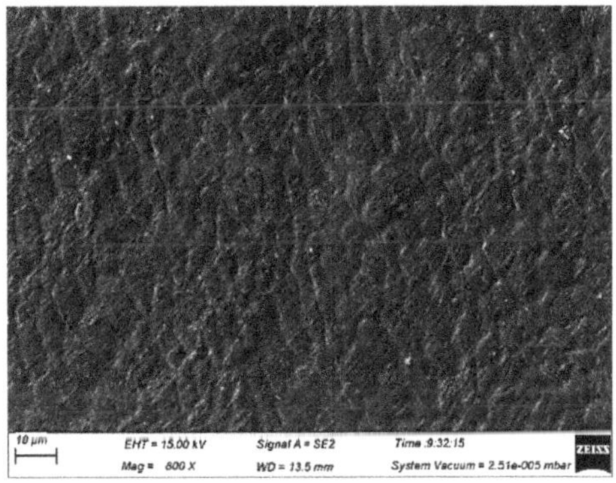

(**d**) The diameter of coarse grains decreased in cross section. (φ3.3592 mm)

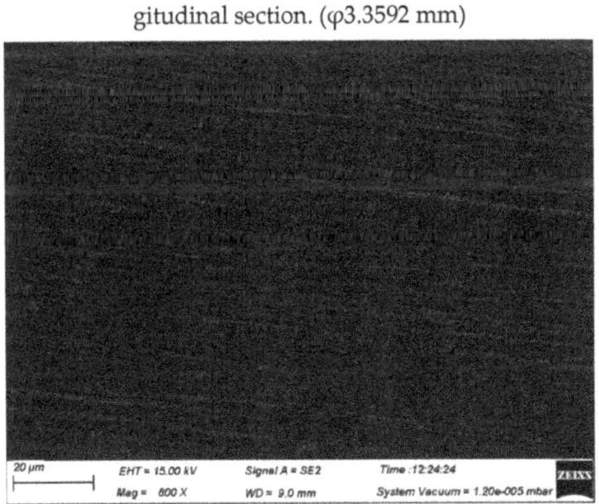

(**e**) The grains become finer and longer in longitudinal section. (φ1.1008 mm)

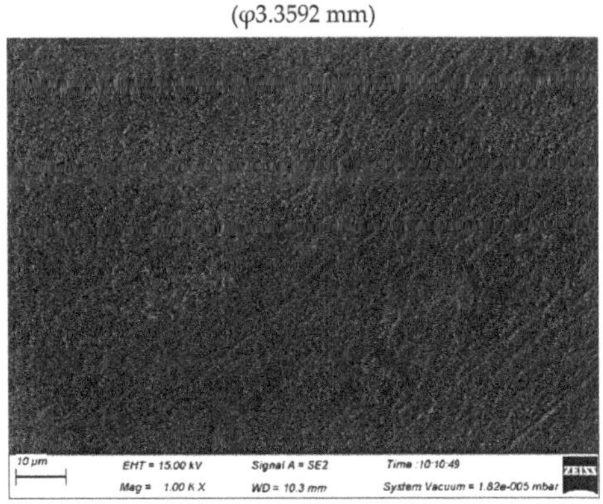

(**f**) The grains diameter becomes smaller in cross section. (φ1.1008 mm)

Figure 2. *Cont.*

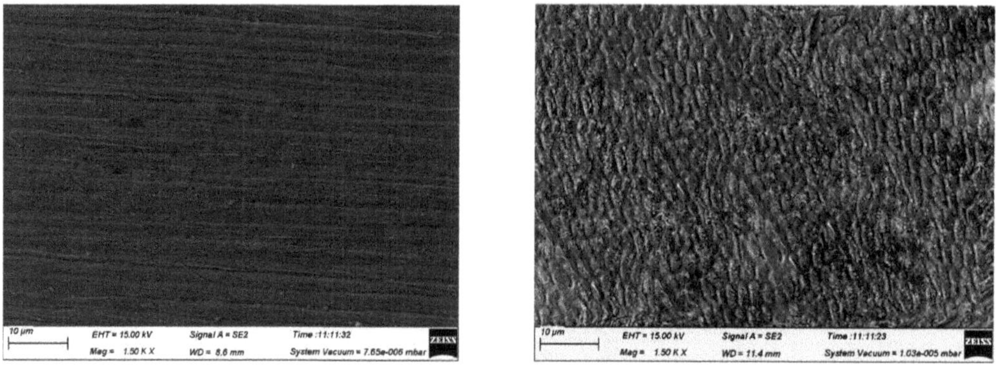

(**g**) The grains are fibrous in longitudinal section. (φ0.2070 mm)

(**h**) The number of grains increases and the diameter decreases in unit area in cross section. (φ0.2070 mm)

Figure 2. Microstructure and morphology of Ag-8.5Au-3.5Pd alloy wires with different deformation variables.

3.2. *Effect of Cold Deformation on Performance of Ag-8.5Au-3.5Pd Alloy Wire*

The curve of mechanical performance of Ag-8.5Au-3.5Pd alloy wires with different deformation values (φ8.0 mm, φ5.2488 mm, φ3.3592 mm, φ2.1499 mm, φ1.1008 mm, φ0.2070 mm) after multi-pass drew are shown in Figure 3 (deformation rates are 0%, 56.95%, 82.37%, 92.78%, 98.11% and 99.93%, respectively). It can be seen that the fracture strength of the wire increases rapidly with the increase in the deformation rate. That is because the alloy wires are processed by cold deformation; the internal grains become finer and smaller and the number of grains increases, as well as the dislocation increasing. However, when cold deformation is processed to a certain extent, the effect of cold deformation on grain refinement is weakened, so when the deformation rate is higher than 92.78%, the increase trend of the strength of the Ag-8.5Au-3.5Pd alloy wire becomes smooth and steady.

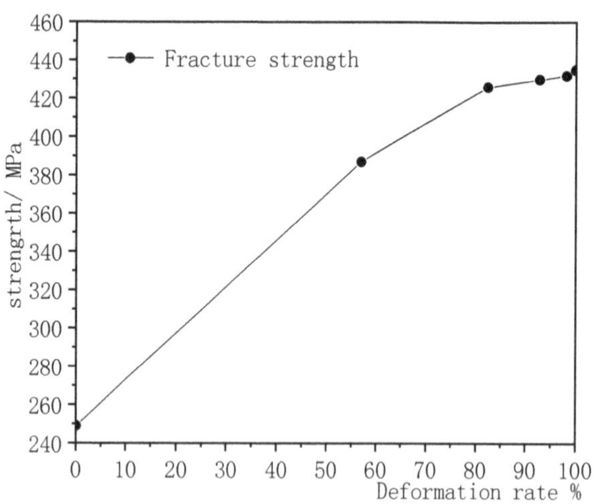

Figure 3. Fracture strength of Ag-8.5Au-3.5Pd alloy wires with different deformation rates.

3.3. *Effect of Different Annealing Temperatures on Electrical Performance of Ag-8.5Au-3.5Pd Alloy Wire*

The curve of the influence at different annealing temperatures on the wire resistivity of φ0.2070 mm Ag-8.5Au-3.5Pd alloy wire is shown in Figure 4. The annealing time is 2.4 s. In this figure, the resistivity decreases from 2.781×10^{-8} Ω·m down to 2.395×10^{-8} Ω·m with the increase in annealing temperature, and the minimum is 2.395×10^{-8} Ω·m when the annealing temperature is 500 °C. The resistivity of the alloy wire almost remains unchanged even if the temperature continues to increase.

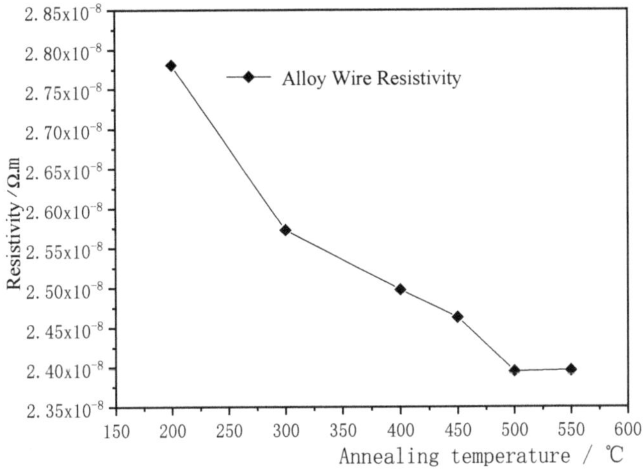

Figure 4. Changes of resistivity of Ag-8.5Au-3.5Pd alloy wire with different annealing temperatures.

The resistivity of metal is mainly caused by phonon, dislocation, point defects (soluble atoms, impurities and vacancies, etc.) and the scattering effect of the interface on electrons [31]. As the increase in the annealing temperature of alloy wires which have been cold worked, the resistivity components change and the corresponding conductivity changes as well. During annealing, some solute atoms gradually precipitate from the matrix under a little free energy released, resulting in a decrease in concentration. Lattice distortion caused by atomic radius difference becomes effectively mitigated and tends towards orderly and periodic arrangement [32,33]. At the same time, the diameter and spacing of the fibers are reduced due to the refinement of the microstructure. The fine fiber structure cannot contain more dislocation substructures, which results in the absorption of dislocations by the grains interface so that the dislocation density decreases. The scattering effect of free electrons during their movement decreased, thus reducing the resistivity of Ag-8.5Au-3.5Pd alloy wires.

3.4. *Effect of Different Annealing Temperatures on the Mechanical Performance of Ag-8.5Au-3.5Pd Alloy Wire*

The influence of different annealing temperatures on the mechanical performance of φ0.2070 mm Ag-8.5Au-3.5Pd alloy wire is shown in Figure 5. It can be seen that the alloy wire begins to recover when the annealing temperature is 300 °C, which eliminates the working-hardening caused by cold working. Its strength decreased significantly from 428 MPa (200 °C) to 307 MPa, and the decrease rate was 28.3%. The elongation increased from 2.2 to 18.5% and the increase rate was 88.1%, and the deformed structure almost all disappeared. When the annealing temperature was 500 °C, the alloy wire fully recovered and began to recrystallize; the grain fiber structure of the alloy wire changed, and the strength decreased to 287 MPa and the elongation increased to 25.7%. As the annealing temperature increased to 550 °C, the grains began to grow, while the strength and elongation decreased to 256 MPa and 20.6%, respectively. It can be seen that the alloy wire has good mechanical performance when the annealing temperature is 500 °C.

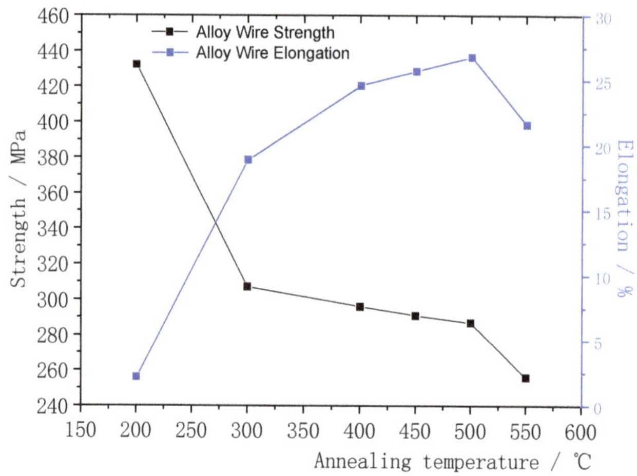

Figure 5. Changes of strength and elongation of Ag-8.5Au-3.5Pd alloy wire with different annealing temperatures.

3.5. Effect of Different Annealing Parameters on Microstructure and Morphology of Ag-8.5Au-3.5Pd Alloy Wire

The microstructure morphology changes of φ0.2070 mm Ag-8.5Au-3.5Pd alloy wire after cold working deformation at different annealing temperatures are shown in Figure 6. When the annealing temperature is 200 °C, only point defects and line defect movements exist in the structure, the lateral displacement of the boundaries of the fiber grain is easy and the interface of the grain's fiber is slightly clear; there is no obvious change in the microstructure in Figure 6a. The point defect and dislocation migrate violently during annealing at 300 °C and 400 °C; the grain's fiber is further separated and coarsened and the deformation structure basically disappears, but most of the dislocation migration is confined to the fiber scale. The fiber shapes observed by SEM are shown in Figure 6b,c. When the annealing temperature is 450 °C, the release tendency of deformation storage energy is increased and significant changes in grain shape are caused by the substantial reduction of crystal defects. A few new large angle grain boundaries are gradually formed, and a few recrystallized grains are produced in the grain's fiber structure. The new nuclei are generally located in the dislocation substructure with a high density, and the deformation structure completely disappeared at this time. The alloy wire is at the initial stage of recrystallization, as shown in Figure 6d. The alloy wire begins to recrystallize when the annealing temperature is 500 °C, as shown in Figure 6e. The grain's fiber is further coarsened locally and evolved into short bar or equiaxed fine grains distributed along the direction of the original fiber. The grains' interface migration is thoroughly activated due to the increase in recrystallized grains and the increase of grain size. In order to reduce the total interfacial energy, the grain boundary migration starts to break through the limit of original grains boundary, which caused obvious distortion and local shrinkage of the grains interface; the fibers are even partially broken and separated, which leads to the gradual disappearance of composite fibers. In Figure 6f, when the annealing temperature was 550 °C, the grains began to grow bigger.

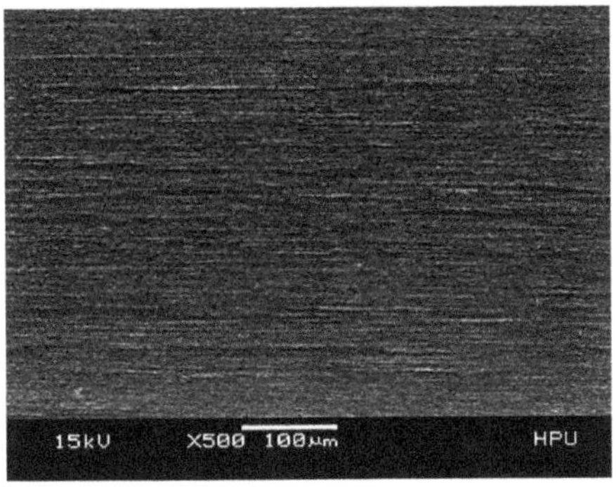

(**a**) The annealing temperature is 200 °C, microstructure change is not obvious.

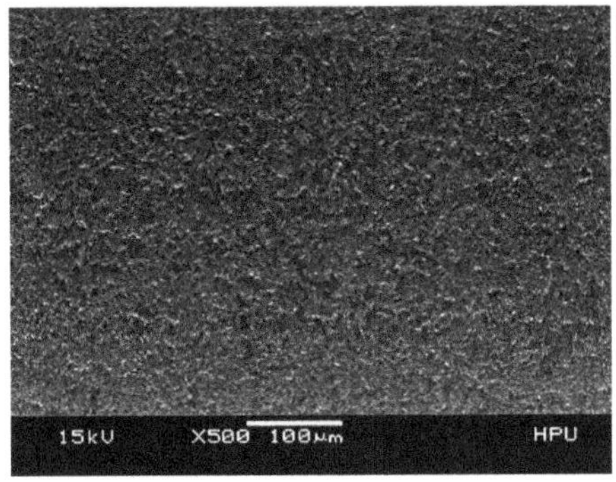

(**b**) The annealing temperature is 300 °C, the grains fiber begins to separate.

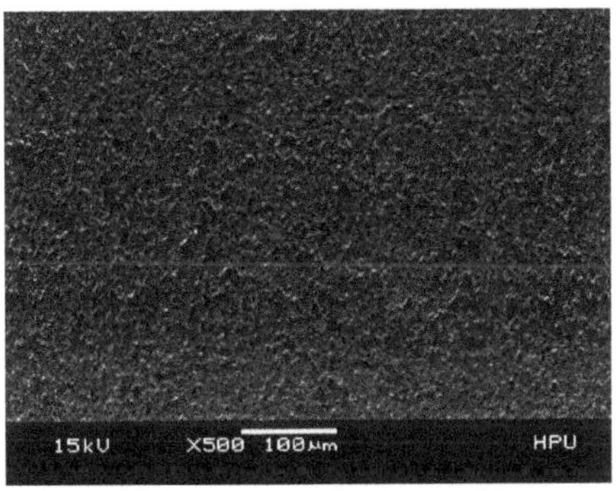

(**c**) The annealing temperature is 400 °C, the deformed tissue almost disappeared.

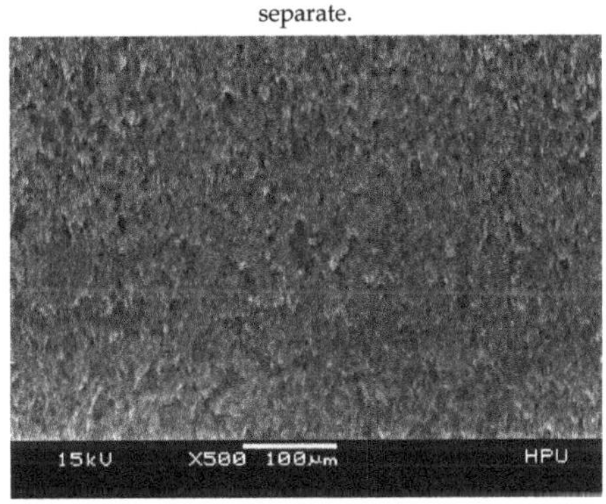

(**d**) The annealing temperature is 450 °C, a few new grain boundaries are formed.

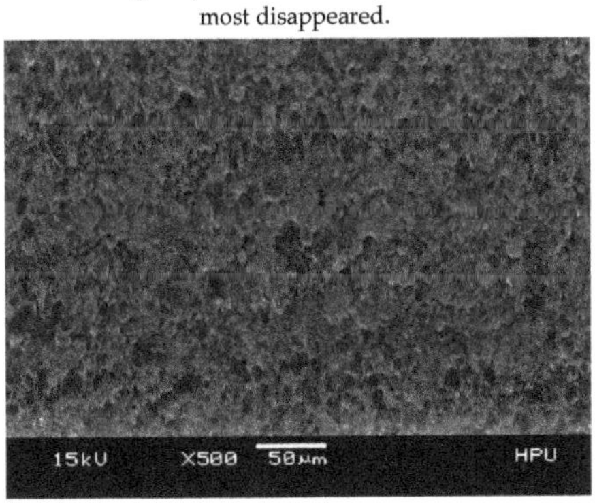

(**e**) The annealing temperature is 500 °C, the alloy wire begins to recrystallize.

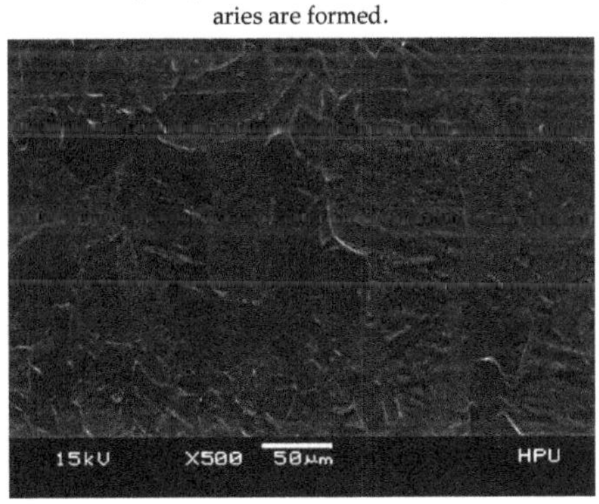

(**f**) The annealing temperature is 550 °C, the grain size becomes coarse.

Figure 6. Microstructure and morphology of Ag-8.5Au-3.5Pd alloy wire at different annealing temperatures.

3.6. Effect of Wire Drawing on Fine Ag-8.5Au-3.5Pd Alloy Wire

The guide wheel quality has a great influence on the surface of the Ag-8.5Au-3.5Pd alloy wire in the fine drawing process. The surface damage can be caused by the inflexible rotation of the guide wheel or surface defect of the guide wheel during wire drawing, as shown in Figure 7. During the drawing process, stress will be concentrated on the damage or scratch on the surface of the alloy wire, which can easily cause serious distortion or fracture, while the contamination on the surface increases the pulling tension during the drawing process, resulting in uneven size of the alloy wire diameter.

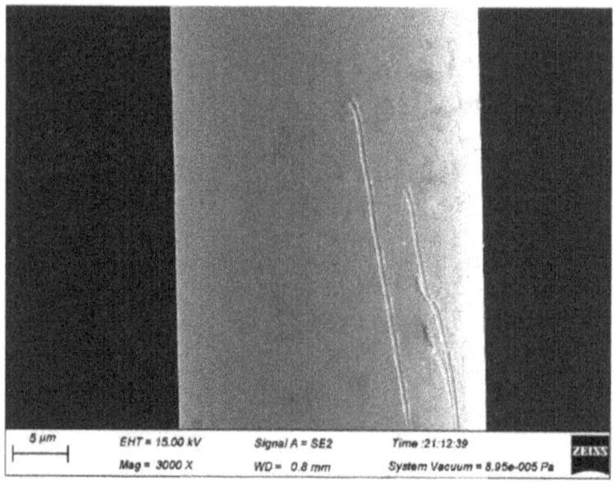

Figure 7. Scratches on the surface of Ag-8.5Au-3.5Pd alloy wire due to dirty guide wheel (0.020 mm).

The concentration of lubricant has a serious influence on the wire drawing process during the fine machining of the Ag-8.5Au-3.5Pd alloy wire. Low concentration solution will lead to poor lubrication, increase wear of the die and cause scratches, grooves and other defects on the surface of the alloy wire, while high concentration solution will block the hole and increase the drawing force. When the lubricant concentration is too high, the lubricant will stay in the lubrication zone of the die, which prevents the lubricant from entering the sizing and compression zones, so that the lubricant cannot play a lubricating role in processing. This causes the pulling force to increase and the wire to become finer, even causing defects and lubricant residue on the wire surface. Residual lubricant on the wire surface will cause serious contamination and accelerated wear of the dies. Water-soluble lubricant is used in the drawing process of alloy wire and the concentration of lubricant is 0.5%. The dilution of lubricant uses deionized water, because common tap water contains a lot of calcium ions, which are easy to alkalize and difficult to clean. This will have a negative impact on the surface of the alloy wire.

The surface quality of the Ag-8.5Au-3.5Pd alloy wire is highly required. If contamination exists on the wire surface (as shown in Figure 8), the bonding strength will be reduced, especially during the second bonding point process. Therefore, the surface of Ag-8.5Au-3.5Pd alloy wire needs to be cleaned after fine wire drawing. The cleaning method is on-line ultrasonic cleaning. Figure 9 shows the high-quality surface of Ag-8.5Au-3.5Pd alloy wire after cleaning.

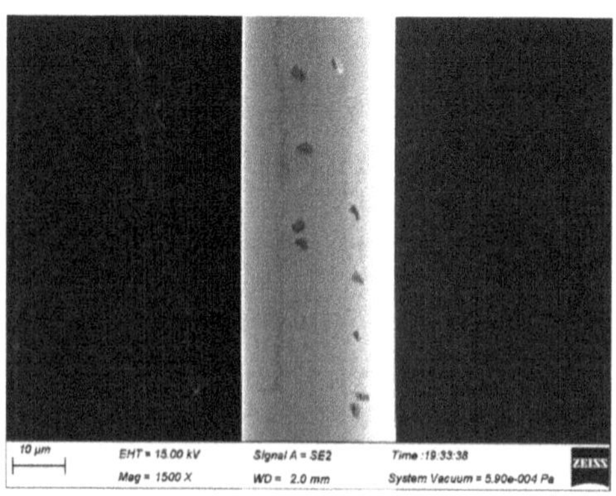

Figure 8. Surface contamination of Ag-8.5Au-3.5Pd alloy wire due to dirty guide wheel (0.020 mm).

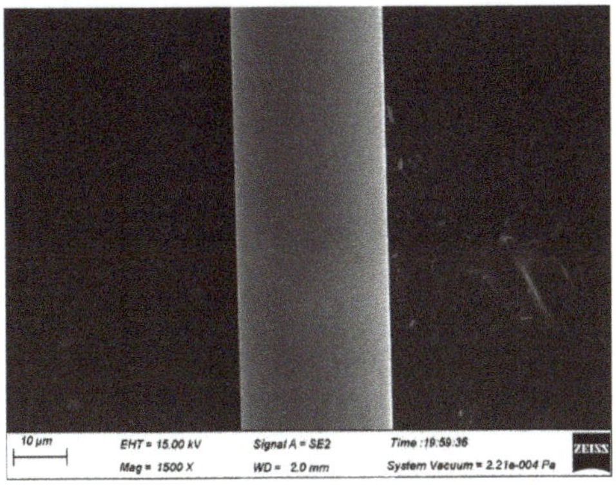

Figure 9. Perfect Surface of Ag-8.5Au-3.5Pd alloy wire (φ0.025 mm).

3.7. *Effect of Annealing Temperature on Mechanical Performance of Fine Ag-8.5Au-3.5Pd Alloy Wire*

The φ0.020 mm Ag-8.5Au-3.5Pd alloy wire was heat treated at different temperatures (the annealing time and speed were 0.8 s and 90 m/min, respectively). The influence of different annealing temperatures on the fracture force and elongation of the alloy wire was studied, and the results are shown in Figure 10.

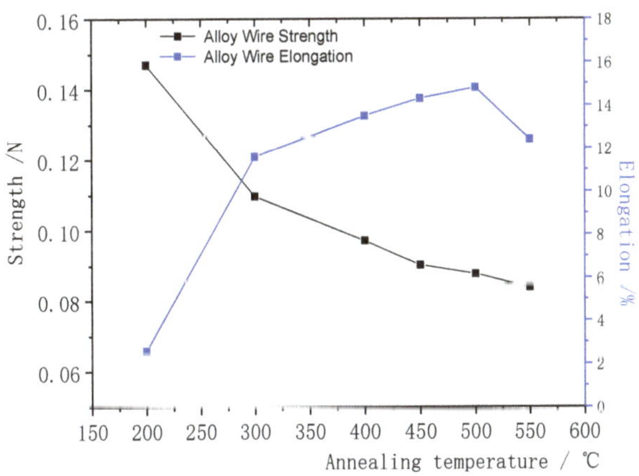

Figure 10. Relationship between annealing temperature and fracture strength and elongation of Ag-8.5Au-3.5Pd alloy wires.

It can be seen that recovery occurred in the Ag-8.5Au-3.5Pd alloy wire with the increase in the annealing temperature, and eliminated the work hardening caused by the cold working; its tensile fracture force reduced while elongation increased. When the annealing temperature increased to 300 °C, the degree of recovery was further strengthened, the fracture force decreased smoothly to 0.1097 N and the elongation increased to 11.6%. The fracture force of the alloy wire decreased rapidly to 0.0971 N while the elongation increased to 13.5%, with the annealing temperature increasing to 400 °C. The recovery process is completed at this stage. When the annealing temperature was 450 °C, the fracture force of the Ag-8.5Au-3.5Pd alloy wire decreased to 0.0902 N; at the same time, the elongation increased to 14.3%. When the annealing temperature was 500 °C, the fracture force was 0.0877 N and the elongation was 14.8%; the alloy wire has excellent mechanical performance in this case. The Ag-8.5Au-3.5Pd alloy wire begins to recrystallize and produces coarse grains with further increase in the annealing temperature (550 °C), as shown in Figure 11. Coarse grains reduce the hindrance to dislocation movement, so that the alloy wire is more prone to stress concentration when subjected to tension, which results in a decrease in mechanical performance. The tensile force decreased to 0.0839 N and the elongation decreased to 12.4%. It can be seen from Figure 10 that the Ag-8.5Au-3.5Pd alloy wire has excellent mechanical performance when the annealing temperature is 500 °C.

Figure 11. Coarse grains structural of Ag-8.5Au-3.5Pd alloy wire at 550 °C annealing temperature.

3.8. Effect of Annealing Tension on Surface Quality of Ag-8.5Au-3.5Pd Alloy Wire

The φ0.020 mm Ag-8.5Au-3.5Pd alloy wire was heat treated in different tension values (1.5 g, 2.0 g, 2.5 g, 3.0 g, 3.5 g). The annealing time, speed and temperature are 0.8 s, 90 m/min and 500 °C, respectively.

When the annealing tension is small (1.5 g, 2.0 g), the alloy wire will vibrate seriously in the annealing tube, which results in contact with the annealing tube wall. The surface of the alloy wire is burned (as shown in Figure 12) because the wall temperature of the annealing tube is higher than the center temperature. When the annealing tension is 2.5 g or 3.0 g, the surface of the alloy wire is smooth and has no damage, as shown in Figure 13. The dynamic recovery and recrystallizing of the Ag-8.5Au-3.5Pd alloy wire will occur at a high annealing temperature. When the annealing tension is 3.5 g, because the tension acting on the wire is large at this time, the grains will slip under the tension and form corrugated defects on the surface of the wire, as shown in Figure 14. This defect will reduce the strength of the second bonded point, and thus reduce the reliability of the device.

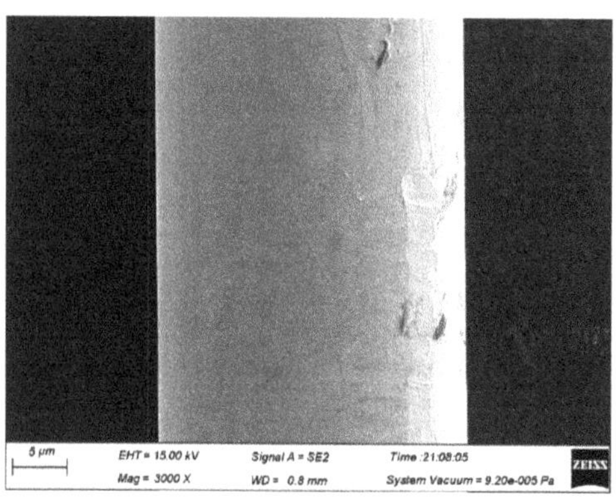

Figure 12. Surface mechanical damage of Ag-8.5Au-3.5Pd alloy wire caused by small tension (1.5 g) after anneal.

Figure 13. Good surface morphology of Ag-8.5Au-3.5Pd alloy wire after annealing (tension 2.0 g or 3.0 g).

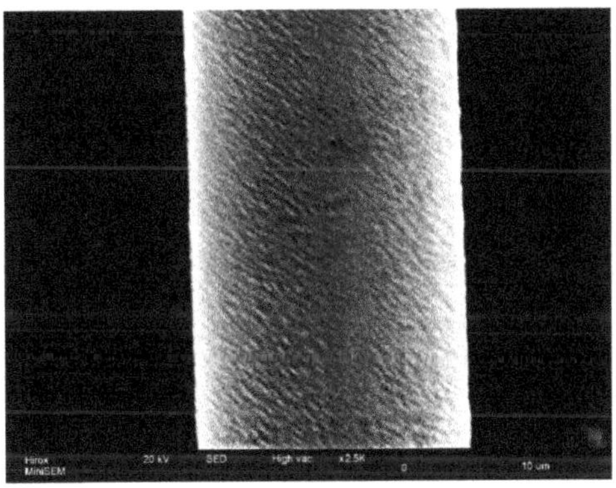

Figure 14. Surface corrugation defects of Ag-8.5Au-3.5Pd alloy wires due to excessive tension (3.5 g) during annealing.

In addition, the Ag-8.5Au-3.5Pd alloy wire will drawn finer in the annealing tube due to excessive tension (3.5 g). After annealing, the diameter of the φ 0.020 mm Ag-8.5Au-3.5Pd alloy wire becomes φ 0.0188 mm and the wire diameter is reduced by 6%, which cannot meet the requirements of industrialization. Therefore, the tension should be in the range of 2.5–3.0 g when the φ0.020 mm Ag-8.5Au-3.5Pd alloy wire is annealing.

4. Conclusions

(1) When Ag-8.5Au-3.5Pd alloy wire is plastic-deformed, metallic grains are gradually straightened and refined and finally show the fibrous structure. With the increase in deformation rate, the strength of the alloy wire increases, and the increase tends to be flat when the deformation rate is higher than 92.78%.

(2) The resistance value of Ag-8.5Au-3.5Pd alloy wire decreases from 2.781×10^{-8} Ω·m to the minimum 2.395×10^{-8} Ω·m (500 °C) with increasing annealing temperature, and it remains constant even if the temperature continues to rise.

(3) With the increase of annealing temperature, the Ag-8.5Au-3.5Pd alloy wire recovers and eliminates the hardening which was caused by cold working, and its strength is significantly reduced. When the annealing temperature is 500 °C, the alloy wire has good mechanical performance, while the grain size begins to grow and its mechanical performance decreases once the annealing temperature is increased to 550 °C.

(4) The φ0.020 mm Ag-8.5Au-3.5Pd alloy wire will shake seriously in the annealing tube if the tension is small (1.5–2.0 g), which would cause burns on the wire surface. When

the tension is large (3.5 g), corrugated defects would appear on the surface of the alloy wire. The alloy wire has a high-quality surface when the tension range is 2.5–3.0 g.

Author Contributions: Conceptualization, J.C. and K.S.; methodology, J.C.; software, G.Y.; validation, J.Z., H.T. and G.Y.; formal analysis, J.C. and J.Z.; investigation, C.C. and Y.G.; resources, J.C.; data curation, J.C.; writing—original draft preparation, J.C.; writing—review and editing, J.Z.; visualization, B.W.; supervision, C.L.; project administration, B.W.; funding acquisition, J.C. All authors have read and agreed to the published version of the manuscript.

Funding: This research was funded by Special Project of Fundamental Scientific Research Funds of Henan Polytechnic University, Fund number: NSFRF210102.

Conflicts of Interest: The authors declare no conflict of interest.

References

1. Chen, Y.; Xie, M.; Wang, S.; Zhang, J.; Yang, Y.; Liu, M.; Wang, S.; Hu, J.; Li, A.; Wei, K. Research Progress of Precious Metal Wire Materials. *Noble Met.* **2014**, *3*, 66–70.
2. Yang, G.; Guo, Y.; Kong, J.; Dao, P.; Guan, W. Effect of Rare Earth Elements on Microstructure and performance of Bond Alloy Wire. *Noble Met.* **2011**, *32*, 16–19.
3. Ding, Y.; Cao, J.; Xu, G.; Co, S.; Hu, Y. Research and Application of Cu Wire for Electronic Packaging. *Cast. Technol.* **2006**, *27*, 971–974.
4. Zhong, M. *First Principle Study on the Effect of Trace Elements on the performance of Copper-based Wires*; Chongqing University of Technology: Chongqing, China, 2019.
5. Feifei, K.; Yongjin, W.; Jianwen, K. Research status of substitute products for gold wires used in microelectronics packaging. *Precious Met.* **2017**, *38*, 81–86. (In Chinese)
6. Wu, J.; Luo, H. The developments and challenges of Cu wire. *Electron. Prod. Reliab. Environ. Test.* **2008**, *26*, 39–42. (In Chinese)
7. Toyozawa, K.; Fujita, K.; Minamide, S.; Maeda, T. Development of copper wire application technology. *IEEE Trans. Compon. Hybrids Manuf. Technol.* **1990**, *13*, 667–672. [CrossRef]
8. Shuang, L.; Fuxiang, H.; Cheng, P.; Mingjun, Z.; Baoan, W.; Huiyi, T. Research Progress of Copper and Silver Wires for Electronic Packaging. *Funct. Mater.* **2019**, *50*, 5048–5053+5063.
9. Hong, S.; Hang, C.; Wang, C. Research progress of copper wire ball welding technology. *Mater. Sci. Technol.* **2007**, *5*, 673–678.
10. Cao, J.; Ding, Y.T.; Guo, T.B. Effect of copper properties and bonding parameters on bonding quality. *Mater. Sci. Technol.* **2012**, *20*, 76–79.
11. Cao, J.; Wu, W.X. Effects of Au coated Ag bonding wire properties on bonded quality. *Mater. Sci. Technol.* **2018**, *26*, 96–100.
12. Czerny, B.; Mazloum-Nejadari, A.; Khatibi, G.; Weiss, L.; Zehetbauer, M. Fatigue testing method for finebond wires in an LQFP package. *Microelectron. Reliab.* **2016**, *64*, 270–275. [CrossRef]
13. Gross, D.; Haag, S.; Reinold, M.; Schneider-Ramelow, M.; Lang, K.D. Heavy copper wire-bonding on silicon chips with aluminum-passivated Cu bond-pads. *Microelectron. Eng.* **2016**, *156*, 41–45. [CrossRef]
14. Ma, X.; Fan, H.; Li, S.; Li, Y.; Du, J. Development of non oxidation bond alloy silver alloy wire. *Gold* **2016**, *63*, 336–341.
15. Qin, W.; Cohen, I.M.; Ayyaswamy, P.S. Ball size and HAZ as Functions of EFO parameters for gold Bonding wire. *Adv. Electron. Packag.* **1997**, *1*, 391–398.
16. Cao, J.; Fan, J.; Gao, W. Effect of Ag-10Au-3.6Pd alloy bonding wire properties and structure on bond strengths and reliability. *Intern. Conf. Electron. Packag.* **2016**, *4*, 82–87.
17. Hang, C.J.; Song, W.H.; Lum, I.; Mayer, M.; Zhou, Y.; Wang, C.Q.; Moon, T.J.; Persic, J. Effect of electronic flame off parameters on copper bonding wire: Free-air Ball deformability, heat affected zone length, heat affected zone breaking force. *Microelectron. Eng.* **2009**, *86*, 2094–2103. [CrossRef]
18. Chen, J.; Lai, Y.S.; Wang, Y.W.; Kao, C.R. Investigation of growth behavior of Al–Cu intermetallic compounds in Cu wire. *Microelectron. Reliab.* **2011**, *51*, 125–129. [CrossRef]
19. Kunimune, T.; Kuramoto, M.; Ogawa, S.; Sugahara, T.; Nagao, S.; Suganuma, K. Ultra thermal stability of LED die-attach achieved by pressureless Ag stress-migration at low temperature. *Acta Mater.* **2015**, *89*, 133–140. [CrossRef]
20. Schneider-Ramelow, M.; Ehrhardt, C. The reliability of wire using Ag and Al. *Microelectron. Reliab.* **2016**, *63*, 336–341. [CrossRef]
21. Schneider-Ramelow, M.; Geißler, U.; Schmitz, S.; Grübl, W.; Schuch, B. Development and Status of Cu Ball/Wedge in 2012. *J. Electron. Mater.* **2013**, *42*, 558–595. [CrossRef]
22. Tseng, Y.; Hung, F.; Lui, T. Microstructure, tensile and electrical performance of gold-coated silver wire. *Microelectron. Reliab.* **2015**, *55*, 608–612. [CrossRef]
23. Hsueh, H.W.; Hung, F.Y.; Lui, T.S.; Chen, L.H.; Chang, J.K. Microstructure, electric flame off (EFO) characteristics and tensile performance of silver-lanthanum alloy wire. *Microelectron. Reliab.* **2014**, *54*, 2564–2569. [CrossRef]
24. Li, B.H.; Ye, J.T.; Liao, J.K.; Zhuang, P.L.; Zhang, Y.P.; Li, J.Y. Effect of pretreatments on the metal–ceramic strength of a Pd–Ag alloy. *J. Dent.* **2014**, *42*, 319–328. [CrossRef]

25. Huang, W.H.; Lin, K.L.; Lin, Y.W.; Cheng, Y.K. The Intermetallic Compound Formation for the Wire Bond Between an Al Pad and Ag-xPd Alloy Wire. *J. Electron. Mater.* **2016**, *45*, 6130–6136. [CrossRef]
26. Chuang, T.; Chen, C. Mechanism of the Electromigration in Ag-Pd Alloy Wires. *Metall. Mater. Trans.* **2018**, *49*, 5904–5910. [CrossRef]
27. Feng, D.; Taskinen, P. Thermodynamic performance of silver–palladium alloys determined by a solid state electrochemical method. *J. Mater. Sci.* **2014**, *49*, 5790–5798. [CrossRef]
28. Guo, R.; Hang, T.; Mao, D.; Li, M.; Qian, K.; Lv, Z.; Chiu, H. Behavior of intermetallics formation and evolution in Ag-8Au-3Pd alloy wire bonds. *J. Alloy. Compd.* **2014**, *588*, 622–627. [CrossRef]
29. Cao, J.; Fan, J.; Gao, W. Effects of Drawing and Annealing on performance of Ag-4Pd Alloy Wire. *J. Mech. Eng.* **2016**, *52*, 92–97. [CrossRef]
30. Cao, J.; Fan, J.; Gao, W. Effects of performance and Structure of Ag-4Pd Alloy Wire on Strength. *Rare Met. Mater. Eng.* **2018**, *47*, 1836–1841.
31. Hong, S.; Hill, M. Mechanical stability and electrical conductivity of Cu-Ag filamentary microcomposites. *Mater. Sci. Eng. A* **1999**, *264*, 151–158. [CrossRef]
32. Ohsaki, S.; Yamazaki, K.; Hono, K. Alloying of immiscible grains in wire drawn Cu-Ag filamentary composites. *Scr. Mater.* **2003**, *48*, 1569–1574. [CrossRef]
33. Heringhaus, F.; Schneider-Muntau, H.J.; Gottstein, G. Analytical modeling of the electrical conductivity of metal matrix composites: Application to Ag-Cu and Cu-Nb. *Mater. Sci. Eng. A* **2003**, *347*, 9–20. [CrossRef]

micromachines

MDPI

Article

Fabrication of Needle-like Silicon Nanowires by Using a Nanoparticles-Assisted Bosch Process for Both High Hydrophobicity and Anti-Reflection

Zengxing Zhang [1], Guohua Liu [2] and Kaiying Wang [1,*]

[1] Department of microsystems, University of South-Eastern Norway, 3184 Horten, Norway; zengxingzhang88@163.com
[2] Beijing Key Laboratory of Multiphase Flow and Heat Transfer for Low Grade Energy Utilization, North China Electric Power University, Beijing 102206, China; liuguohua126@126.com
* Correspondence: kaiying.wang@usn.no

Abstract: In this work, a modified Bosch etching process is developed to create silicon nanowires. Au nanoparticles (NPs) formed by magnetron sputtering film deposition and thermal annealing were employed as the hard mask to achieve controllable density and high aspect ratios. Such silicon nanowire exhibits the excellent anti-reflection ability of a reflectance value of below 2% within a broad light wave range between 220 and 1100 nm. In addition, Au NPs-induced surface plasmons significantly enhance the near-unity anti-reflection characteristics, achieving a reflectance below 3% within the wavelength range of 220 to 2600 nm. Furthermore, the nanowire array exhibits super-hydrophobic behavior with a contact angle over ~165.6° without enforcing any hydrophobic chemical treatment. Such behavior yields in water droplets bouncing off the surface many times. These properties render this silicon nanowire attractive for applications such as photothermal, photocatalysis, supercapacitor, and microfluidics.

Keywords: silicon nanowires; nanoparticles; Bosch process; anti-reflection; hydrophobicity

Citation: Zhang, Z.; Liu, G.; Wang, K. Fabrication of Needle-like Silicon Nanowires by Using a Nanoparticles-Assisted Bosch Process for Both High Hydrophobicity and Anti-Reflection. *Micromachines* **2021**, *12*, 1009. https://doi.org/10.3390/mi12091009

Academic Editor: Lucia Romano

Received: 31 July 2021
Accepted: 23 August 2021
Published: 25 August 2021

1. Introduction

Silicon nanowires have been widely used over the past years in many emerging fields, including photodetection [1,2], photocatalysis [3,4], thermoelectric [5,6], and quantum information processing [7,8]. Nanowires are one-dimensional materials that present the shape of needles, rods, or pillars [9,10]. Their diameter lies in the range of nanometers to a few hundred nanometers [11], while their height can be as large as several micrometers [12]. Apart from the bottom-growth approaches, various patterning and dry etching processes have also been developed to realize nanowire fabrication with semiconductors [13–15]. Interestingly, most of the reported nanowire arrays are patterned by using electron-beam and holographic lithography techniques [16], which are attractive in generating a periodic pattern with a high degree of uniformity. However, the approaches mentioned above are time-consuming for large area nanowire fabrication and exhibit low throughput for mass production. As far as dry etching is concerned, it generally involves mask materials such as photoresist, silicon dioxide, and Teflon microspheres to fabricate structures with high aspect ratio profiles [16]. For instance, the employment of metal masks requires the enforcement of an additional lift-off treatment that decreases the pattern resolution and increases the process complexity. As a result, a specific eliminating step is needed that can cause irreducible contamination and damage. Recently, a self-assembled mask approach has emerged that is expected to address the issues mentioned above [17,18]. The method refers to forming a layer of gold NPs, which is formed by depositing a thin gold film. The subsequent application of a thermal annealing treatment to aggregate to be a particle-like hard mask without enforcing any other lithography procedures.

The frequently used etching treatments include laser ablation [19], metal-catalytic chemical etching [20,21], and reactive ion etching (RIE) [12,22]. Laser ablation is a random etching that can cause undercutting of the mask. The silicon wafer's crystal orientation binds the catalyst-assisted etching. In contrast, deep reactive ion etching (DRIE) is the highly anisotropic etching method used to create holes and trenches with high aspect ratios and close-to-vertical sidewalls on various silicon substrates [17]. The Bosch process [23–25] is named after the invention of two employees of the German Robert Bosch company and its name originates from the company name. This process is also known as pulsed- or time-multiplexed-etching and includes the cycle between SF_6 etching and C_4F_8 passivation, each phase lasting a few milliseconds. The passivation layer protects the hole/trench sidewalls from excessive bombardment. At the same time, the directional fluorine/sulfur ions bombard and remove the bottom passivation layer faster than that along the sidewalls. Such etch/protect steps alternate in sequence resulting in minimal isotropic etch taking place only at the bottom of the etched patterns, which makes the Bosch process become a tremendous potential method in nanostructure fabrication.

A novel one-step Bosch process is presented in this work to fabricate silicon nanowires with controllable surface density and a high aspect ratio. Our approach contains three main processing steps: NPs deposition as a mask material, modified Bosch etching, and removal of the mask. It is also attainable to change the nanowire's structural profile by tuning the mask size and total etching duration. The modified Bosch process can transfer the NPs-based mask pattern into the silicon substrate and remove the particles in the etching process. In addition, the NPs-based mask is created by employing the magnetron sputtering for the deposition of a thin gold film (thickness from 4 to 10 nm) and subsequent thermal annealing at 500 °C for 1 h. The annealing temperature, holding time, and film thickness are crucial parameters to control the particle surface density and size. As a result, the enforcement of a higher annealing temperature, elevated holding times, and the employment of a thicker film led to a low surface density of the NPs and decreased filling density of the nanowires at last. In addition, our unique silicon nanowire arrays exhibit extremely superhydrophobic behavior with a contact angle over ~165.6° without any hydrophobic chemical treatment. Such behavior permits the water droplets to bounce off the surface many times, making it extremely difficult to conduct pendant-drop measurements. Besides, the fabricated silicon nanowire structure displays lower than 2% near-unity reflectance properties within an optical wave range of 220 to 1100 nm. It is interesting to notice that after incorporating the Au NPs and inducing the surface plasmons effect on their surface, a near-unity reflectance factor below 3% in the broadband range of 220 to 2600 nm is achieved.

2. Materials and Methods

2.1. Fabrication Process

The involved fabrication processes are schematically illustrated in Figure 1. Firstly, polished (100)-oriented p-type silicon wafers with a resistivity of 1–20 Ω·cm and thickness of 500 ± 10 μm were selected to fabricate the silicon nanowires. Next, all wafers were cleaned using a buffered oxide etch (BOE) solution at room temperature, followed by acetone, isopropanol, and deionized water (DI water) rinsing to remove the native oxide, residuals, and organic contamination. Next, gold thin films with various thicknesses of 4, 6, 8, and 10 nm were deposited on the wafer surface by employing the magnetron sputtering technique at room temperature (DC sputtering system, AJA International, Scituate, MA, USA). Next, the Au NPs-based masks were created by applying a rapid thermal annealing step at 500 °C for one hour. The Au NPs size presents a strong dependence on the film thickness. Subsequently, the silicon wafer was DRIE-etched with a modified Bosch process to create the nanowires. The etching process was conducted within an RIE cluster reactor (Plasma Pro100 Cluster, Oxford Instruments, Abingdon, UK).

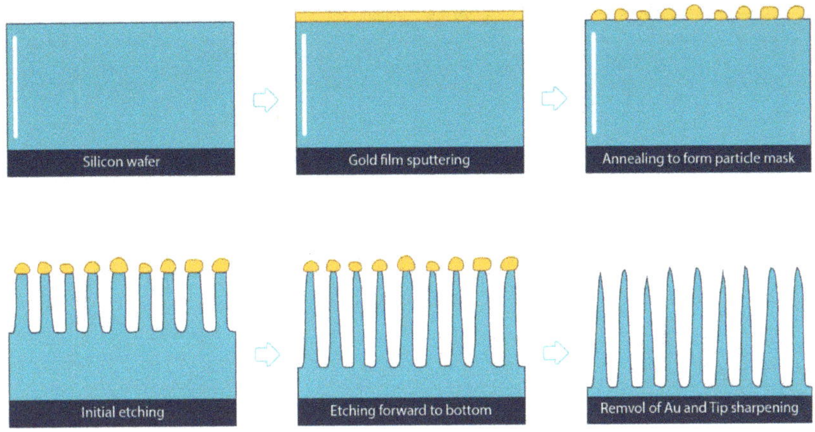

Figure 1. The schematic diagram of the nanowire fabrication process. The gold thin film is deposited on the P-type silicon wafer. Thermal annealing is employed to create gold particle mask. The followed etching treatment consists of initial Bosch etching, forward etching for more undercut, and the continue etching to remove Au nanoparticles.

2.2. Etching Conditions

The Bosch process is carried out through the fast alternative between two modes, namely the bombardment reaction by SF_6 plasma and the protection by polymer C_4F_8 to obtain structures with nearly vertical sidewalls. Table 1 illustrates the modified etching recipe based on the Bosch process. The execution of such etching experiments requires the utilization of a helicon antenna-coupled plasma reactor that operates in the inductive mode. With the assistance of turbine vacuum pumps, we can rapidly and precisely adjust various experimental parameters such as gas flux, pressure, and plasma density. The employed single etching loop mainly consists of the three following steps: protection polymer deposition, bottom polymer breakthrough, and depth direction etching. Initially, the protection deposition step is conducted by introducing repeatable flows of C_4F_8 up to the values of 60 SCCM to form a thick polymer layer on both the metal NPs and the pore sidewall. Such thick polymer is wrapped on the metal NPs, which is beneficial to avoid the mask undercut and achieve a high selection ratio between the particle mask and the silicon. For the subsequent deposition step, a C_4F_8 flow rate of 5 SCCM is selected to form a thin protection polymer at the bottom of the pore. In the meantime, the SF_6 flow rate of 160 SCCM is set to avoid fast C_4F_8 deposition to block the nanowires' interval spacing. Such balance between the deposit and the etching procedures is adjustable according to the structure pitch size. Next, the bottom breakthrough step aims to dissolve the protective polymer and expose the silicon substrate at the pore bottom. Finally, the depth-directional etching occurs, where the posterior region of the silicon reacts with SF_6 faster than that on the sidewall, thus realizing a highly anisotropic etching. Besides, a silicon wafer with Au film deposited on its top surface is also selected as the platform to support the silicon pieces to be etched. The platform temperature is retained constant at the value of 5 Celsius by flowing helium gas. We also used the highly thermal conductivity silicone grease to bond the etching pieces on the table silicon wafer. The goal was to accurately control the uniform piece temperature distribution and to be consistent with the set value of the platform temperature.

Table 1. Nanoparticles-assisted Bosch process.

Main Steps in 1 Etching Loop	SF_6 Gas Flow (sccm)	C_4F_8 Gas Flow (sccm)	ICP Power (W)	HF Power (W)	O_2 Gas Flow (sccm)	Table Temperature (°C)	Pressure (mTorr)	Helium Backing (Torr)	Step Time (ms)
Pre-deposition	10	200	1500	5	0	5	0	10	25
Deposition	5	60	1250	5	0	5	20	10	550
Deposition Sub 1	20	60	1250	5	0	5	20	10	50
Deposition Sub 2	160	60	1250	5	0	5	30	10	100
Deposition Sub 3	160	5	2000	5	0	5	30	10	50
Deposition Sub 4	160	5	2000	60	0	5	30	10	50
Breakthrough	200	5	2000	60	0	5	30	10	325
Breakthrough sub 1	200	5	2000	60	0	5	30	10	100
Breakthrough Sub 2	200	5	2500	60	0	5	80	10	50
Breakthrough Sub 3	200	5	2500	0	0	5	80	10	50
Etch	500	5	2500	0	0	5	80	10	600
Etch Sub 1	1	120	2500	0	0	5	80	10	150
Etch Sub 2	5	120	1250	0	0	5	20	10	100

2.3. Surface Characterization

The surface texture morphology was investigated using a scanning electron microscope (SEM, Hitachi UHR FE-SEM SU8230, Hitachi, Tokyo, Japan). The reflectance and transmittance spectra were measured using a UV-Vis-NIR spectrophotometer (Shimadzu UV 3600 plus, Shimadzu, Kyoto, Japan). The $BaSO_4$ compound was used as the reflectance spectrum reference from the nanowire surface, while the aluminum mirror was the reference for reflectance spectra of the polished silicon wafer. The total reflection, diffuse, and mirror reflection were measured in the integrating sphere. Furthermore, the contact angle of the nanowire texture was characterized by using the contact angle meter, which is equipped with a rotatable substrate holder and automated dispenser (CAM-01A, GLOBAL ANALYTICAL, Ankara, Turkey). The mechanical stability experiment was performed by using an automatic dripping device. Five thousand water droplets of twenty microliters were dripped continuously from a distance of 30 cm to strike the silicon surface. The bouncing effect of the water droplet on such a surface was captured and recorded by employing a high-speed video camera (5KF10, Fuhuang Agile Device, Hefei, China).

3. Results and Discussion

Figure 2a divulges the digital photo of the samples with nanowires formed by applying 120 Bosch etching loops. The NPs were created by enforcing an annealing step on a 4 nm Au thick film and served the role of the etching mask. The sample surface exhibits a black visual due to the nanowires' "light trapping" ability [26,27]. This ability originates from the energy decay in the surface light reflections and the optical path length extension through the internal refraction and transmission processes [26,28]. The surface reflection loss stems from both the mirror reflection effect at the smooth substrate and diffusion reflection caused by structures with small aspect ratios. As a result, the employment of surface textures with a configuration of the pore, tip, and deep holes can effectively suppress such reflection losses by permitting more incident light to enter the structure and multi-reflect along the depth direction. As far as the optical path length is concerned, it is significantly larger than the actual substrate thickness. More specifically, it is defined as the distance that an unabsorbed photon could travel within the silicon bulk before it escapes, which implies that light waves will have more opportunities to achieve a coupling decay with silicon [29]. The employment of a structure that induces multiple internal reflections, light refraction, and transmission effects can facilitate such a coupling decay process [30]. As a result, surface structures with a smaller smooth area on top and larger aspect ratios can trap an enhanced incident light, leading to constant attenuation through the consecutive interactions of light and silicon within the structure. Figure 2b illustrates the tilted SEM image of the fabricated nanowire array, showing that the Au NPs are distributed uniformly at the top of the nanowires. It is also indicated that the etching step's depth direction has a high selection ratio between silicon and mask. In addition, the employed etching process is highly anisotropic, which is desired to avoid mask undercut and create structures with elevated aspect ratios. As depicted in Figure 2c, the fabricated nanowires with a height of ~3 μm exhibit a needle-like configuration and have average aspect ratios of about 10:1. Figure 2d divulges the top view image of the silicon nanowires, which indicates a large pore proportion that allows a relatively considerable amount of light to enter and a tiny flat surface proportion on the top to avoid the specular reflection loss of the incident light. In addition, Figure S1 illustrated the digital photographs of a 4 inch wafer of the silicon nanowires formed through 120 etching loops, indicating a good uniformity across the wafer. Table S1 illustrates the etch rate, scallop height, and profile control ability in detail. Figure S2 shows the SEM image at high magnification, which indicates that the sidewalls are smooth and the Bosch scallops have an average height of ~25 nm.

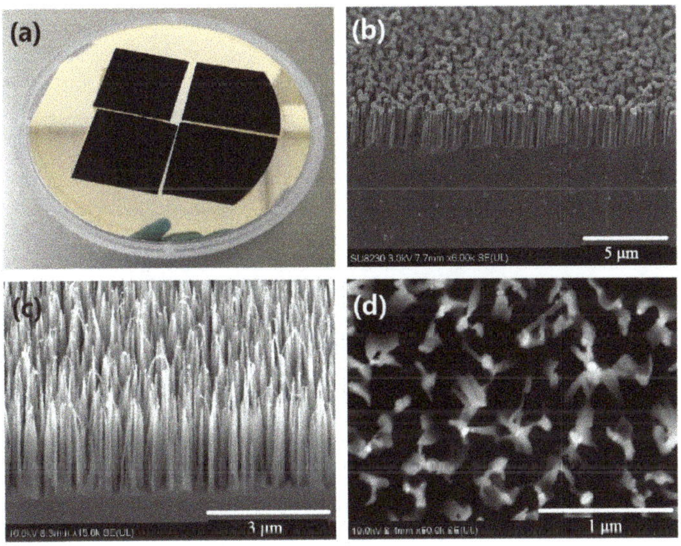

Figure 2. Photograph and SEM images of the silicon nanowires. (**a**) Photograph; (**b**) nanowires with Au nanoparticles on tip top; (**c**) tilted image of the nanowires formed through 120 loops of Bosch process; (**d**) top view of the nanowires formed through 120 loops of Bosch process.

To investigate the impact of the mask size and the duration of the etching process on the silicon nanowire formation, we closely examined the modifications induced by adjusting one etching loop or by employing three different particle sizes. First, Figure S3 illustrates the SEM images of the nanoparticle mask formation before the etching treatment. The Au film with deposition thickness of 4 nm was used, and it presents as continuous islands with spacings. Then, two temperatures of thermal annealing treatment were applied to compare the particle mask formation. It is indicated that the nanowire pitch size mainly depends on the spacing between the isolated particles caused by film thickness and annealing temperature. The nanowires' diameters rely on the randomly formed particles' sizes. Figure 3 illustrates the SEM images of the silicon nanowires formed, enforcing longer etching durations and varying the Au nanoparticle sizes. More specifically, Figure 3a reveals the tilted view and top view images of the nanowire structure formed by applying 160 etching loops and the deposition of a 4 nm thick Au film to create the particle mask. The acquired nanowires exhibited an average height of about 5.62 um, indicating that a longer etching duration produces nanowires with higher heights. Besides,

Figure 3b–d illustrates the tilted and top views images of the silicon nanowires that are etched through the incorporation of particle masks formed by employing 6, 8, and 10 nm thick Au films, respectively. The obtained average nanowire heights are 6.71, 8.83, and 7.78 μm, respectively. The surface density decreases by increasing the particle size, whereas the silicon nanowire formed by 4 nm thick Au film exhibits the highest density. Since a thicker film is more likely to lead to larger NPs, the inter-particle intervals become larger as well. This effect is nicely captured in Figure 3d. Figure S4 illustrates the SEM images at high magnification of the nanowires formed through 160 etching loops; the particle masks are formed by 4, 6, 8, and 10 nm Au film, respectively. It is indicated that the silicon nanowire formed by 4 nm Au film has the highest structure density. The digital images of the nanowire-based samples are presented by employing four different particle masks formed by the deposition of 4, 6, 8, and 10 nm thick Au film. The nanowire surface induced by the presence of 4 nm thick Au film exhibits the darkest appearance, which is obvious through the digital photograph comparison illustrated in Figure S5. Thus, these silicon nanowires were selected for the following test of the optical character.

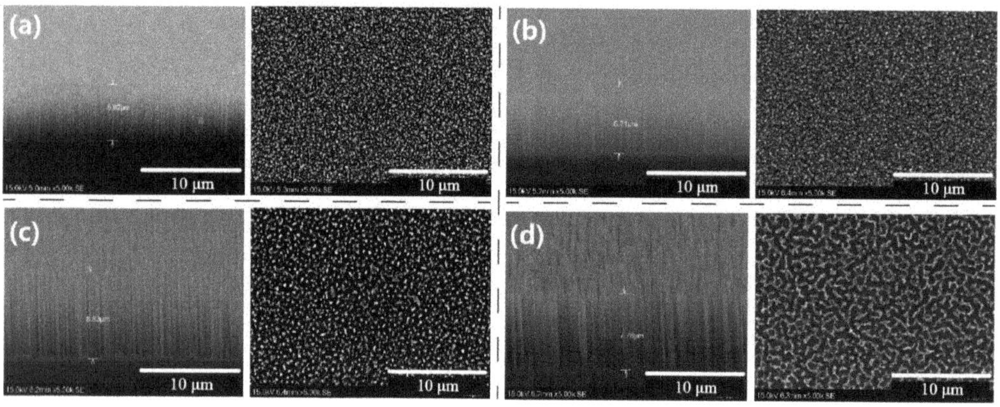

Figure 3. Photograph and SEM images of the silicon nanowires formed through longer etching duration and the varying Au nanoparticle sizes and etching loops. (**a**) Tilted view and top view of the nanowire formed through 160 etching loops, the particle mask is formed by 4 nm Au film; (**b–d**) tilted view and top view of the nanowires formed through 160 etching loops; the particle masks are formed by 6, 8, and 10 nm Au film, respectively.

The optical measurements were carried out to investigate the nanowires' ability to suppress the light reflection. Firstly, a polished silicon wafer was selected as a reference sample for the reflectance and transmittance spectra. Figure 4a demonstrates that the reflectance of the nanowire surface is significantly lower than that of the polished silicon wafer in the wavelength range of 220 to 1100 nm. In addition, the reflectance efficiency of the nanowire sample below the wavelength value of 1000 nm is less than 2%. Figure 4b discloses the comparative spectra, as far as the transmittance property is concerned, indicating that nanowires-based surface exhibits a lower value than the polished one at the wavelength range above 1000 nm. Such a discrepancy could be originated by the change in the material's surface induced by the residuals of the etching reaction. Besides, we also conducted the light reflection comparison of two silicon nanowire samples formed by 4 nm film and two different etching durations. As shown in Figure S6, the silicon nanowire surface formed through 120 etch loops exhibited higher reflection values in the wavelength range of above 1200 nm than that formed through 160 etch loops. In addition, the well-known localized surface plasmon resonance (LSPR) effect [31–33] was applied to decrease the reflectance coupling of the near-infrared light on such nanowires-based surfaces. Au NPs were employed to generate the LSPR. The used LSPR NPs were also loaded via a DC magnetron sputtering deposition of Au, followed by a thermal annealing treatment. As divulged in Figure 4c, the nanowires decorated with Au NPs showed a lower than 3% reflectance at the wavelength range from 220 to 2600 nm.

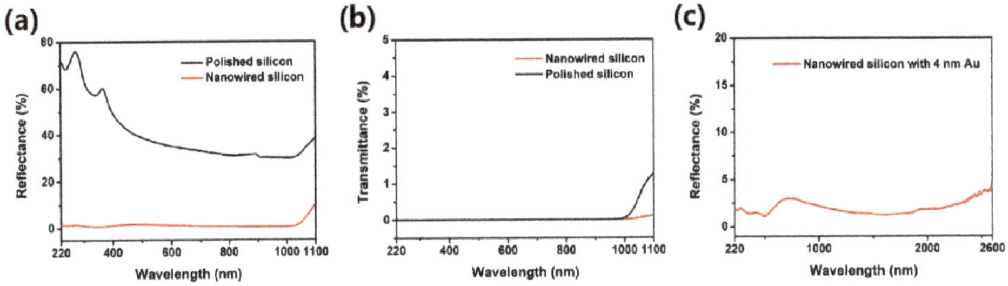

Figure 4. (**a**) The reflectance spectra of the polished silicon surface and the nanowire one; (**b**) the transmittance spectra of the polished silicon surface and the nanowire one; (**c**) the reflectance spectra of the nanowire silicon surface decorated with the Au nanoparticles that are formed by 4 nm Au film.

It is well established that the nanowires are under intensive research for photoelectronic and energy storage-related applications due to their enhanced properties in terms of increased anti-reflectivity, high surface-to-volume ratio, and quantum confinement effects. Besides, silicon nanowires also exhibit unique structural properties that render them quietly different from the bulk silicon. More specifically, they can remarkably enhance the hydrophobicity property of the substrate surface [34,35]. The hydrophobic characteristics are of great importance since they can be utilized in the self-cleaning process, anti-bacterial surfaces, and other biomedical applications. In addition, the octadecyl trichlorosilane (OTS) is frequently applied on nanowire surface as a chemical treatment method to enhance structure-based hydrophobicity. However, such types of treatment tend to increase the contamination levels, requires high cost, and presents poor long-term reliability. It is interesting that our fabricated silicon nanowire arrays exhibited extremely superhydrophobic behavior with a contact angle over the value of ~165.6° and without implementing any hydrophobic chemical treatment, as is revealed in Figure 5a. Such a property permits the water droplet to bounce off the surface and does not allow the pendant-drop measurement to be conducted. Besides, Figure S7 shows the contact angle of the silicon nanowire surface formed by 4 nm Au film and 180 etch loops. Therefore, a hydrophobic-treated micropipette was utilized to enable the execution of the contact angle measurement by releasing a water droplet on such a surface. As illustrated in Figure 5b, the acquired screenshots from a high-speed video (the video frame rate was set to 4000 per second) demonstrate that a water droplet of 1.5 μL drops onto the nanowire surface at the height of 70 mm. Initially, it fully bounces off and consequently continues nine times to rebound along with the vertex's height gradually decreased. The whole process has been recorded by the provided Videos S1 and S2, where the bouncing effect of a water droplet with 1.5 μL on horizontal and inclined nanowire surfaces is presented, respectively. Besides, five thousand water droplets of 20 μL drip continuously struck the silicon nanowires from 30 cm, as depicted in Figure S8, where no apparent damage of the nanowires' surface was detected.

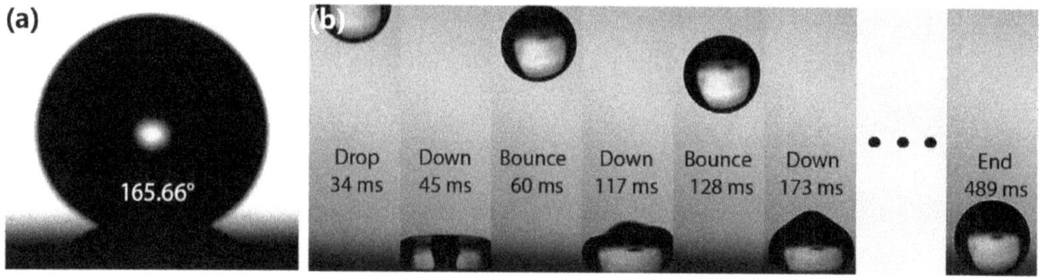

Figure 5. (**a**) Contact angle of the silicon nanowire surface; (**b**) screenshots of a water droplet of 1.5 μL drop onto the surface and bounce ten times, the video frame rate was 4000 per second.

4. Conclusions

A novel DRIE method was demonstrated in this work to create silicon nanowires with controllable surface densities and high aspect ratios. Such a process involves employing an Au NPs-based deposition mask, a modified Bosch etching process, and a mask removal step. As a result, the fabricated silicon nanowire array exhibited an excellent anti-reflection below 2% within a broad light wave range from 220 to 1100 nm. Furthermore, by leveraging the formation of surface plasmons, induced by the creation of Au NPs through the dewetting process of 4 nm thick gold film, a near-unity reflectance of below 3% in the broadband of 220 to 2600 nm was obtained. Besides, our prototypes exhibited extremely super-hydrophobic properties with a contact angle over ~165.6° without applying any hydrophobic chemical treatment, resulting in bouncing off the surface properties of the water droplets. These results render the fabricated silicon nanowire a pretty promising solution for various applications, including developing novel photothermal [5,6,36,37] and photocatalysis processes [38,39], supercapacitor electrodes [40–42], and microfluidic devices [43–45].

Supplementary Materials: The following are available online at https://www.mdpi.com/article/10.3390/mi12091009/s1, Figure S1: Digital photographs of the 4-inch wafer of the silicon nanowires formed through 120 etching loops; Figure S2: The SEM image at high magnification, indicating the smooth sidewalls and the Bosch scallops with the height of ~25 nm; Figure S3: The SEM images of the nanoparticle masks before etching treatment; Figure S4: SEM images at high magnification of the nanowires formed through 160 etching loops, the particle masks are formed by 4, 6, 8, and 10 nm Au film, respectively; Figure S5: Digital photographs of the nanowires formed through 160 etching loops, the particle masks are formed by 4, 6, 8, and 10 nm Au film, respectively; Figure S6: The reflectance spectra of the polished silicon surface and the nanowire ones formed by 4 nm film. The nanowires are formed through 120 and 160 etch loops, respectively, while the planar silicon is used as the reflectance. reference; Figure S7: Contact angle of the silicon nanowire surface formed by 4 nm Au film and 180 etching loops; Figure S8: SEM image of the tilted view of the silicon nanowire array after striking by five thousand water droplets continuously; Table S1: Nanoparticles-assisted Bosch process; Video S1: the whole process of the water droplet with 1.5 μL bouncing on the horizontal nanowire surface; Video S2: the whole process of the water droplet with 1.5 μL bouncing on the inclined nanowire surface.

Author Contributions: Conceptualization, Z.Z. and K.W.; fabrication and methodology, Z.Z. and G.L.; writing—original draft preparation, Z.Z. and K.W.; supervision, K.W.; project administration, K.W.; funding acquisition, K.W. All authors have read and agreed to the published version of the manuscript.

Funding: This research was supported by the KD program of University of South-Eastern Norway, and the Norwegian Micro- and Nano-Fabrication Facility (NorFab, project number 245963/F50), and the Regionale Forskningsfond Vestfold and Telemark (project number 319188).

Data Availability Statement: Data are contained within the article.

Acknowledgments: We thank for Kang Du for assistance in the fabrication experiment.

Conflicts of Interest: The authors declare no conflict of interest.

References

1. Jee, S.-W.; Zhou, K.; Kim, D.-W.; Lee, J.-H. A silicon nanowire photodetector using Au plasmonic nanoantennas. *Nano Converg.* **2014**, *1*, 29. [CrossRef]
2. Soci, C.; Zhang, A.; Bao, X.Y.; Kim, H.; Lo, Y.; Wang, D. Nanowire photodetectors. *J. Nanosci. Nanotechnol.* **2010**, *10*, 1430–1449. [CrossRef] [PubMed]
3. Yu, Y.; Zhang, Z.; Yin, X.; Kvit, A.; Liao, Q.; Kang, Z.; Yan, X.; Zhang, Y.; Wang, X. Enhanced photoelectrochemical efficiency and stability using a conformal TiO_2 film on a black silicon photoanode. *Nat. Energy* **2017**, *2*, 17045. [CrossRef]
4. Alexander, F.; AlMheiri, M.; Dahal, P.; Abed, J.; Rajput, N.S.; Aubry, C.; Viegas, J.; Jouiad, M. Water splitting TiO_2 composite material based on black silicon as an efficient photocatalyst. *Sol. Energy Mater. Sol. Cells* **2018**, *180*, 236–242. [CrossRef]
5. Roder, P.B.; Smith, B.E.; Davis, E.J.; Pauzauskie, P.J. Photothermal heating of nanowires. *J. Phys. Chem. C* **2014**, *118*, 1407–1416. [CrossRef]
6. Chen, R.; Lee, J.; Lee, W.; Li, D. Thermoelectrics of Nanowires. *Chem. Rev.* **2019**, *119*, 9260–9302. [CrossRef] [PubMed]
7. Hocevar, M.; Immink, G.; Verheijen, M.; Akopian, N.; Zwiller, V.; Kouwenhoven, L.; Bakkers, E. Growth and optical properties of axial hybrid III-V/silicon nanowires. *Nat. Commun.* **2012**, *3*, 1266. [CrossRef] [PubMed]
8. Frolov, S.M.; Plissard, S.R.; Nadj-Perge, S.; Kouwenhoven, L.P.; Bakkers, E.P.A.M. Quantum computing based on semiconductor nanowires. *MRS Bull.* **2013**, *38*, 809–815. [CrossRef]
9. Gonchar, K.A.; Kitaeva, V.Y.; Zharik, G.A.; Eliseev, A.A.; Osminkina, L.A. Structural and optical properties of silicon nanowire arrays fabricated by metal assisted chemical etching with ammonium fluoride. *Front. Chem.* **2019**, *6*, 653. [CrossRef]
10. Rashid, J.I.A.; Abdullah, J.; Yusof, N.A.; Hajian, R. The development of silicon nanowire as sensing material and its applications. *J. Nanomater.* **2013**, *2013*, 328093.
11. Huang, Z.; Geyer, N.; Werner, P.; De Boor, J.; Gösele, U. Metal-assisted chemical etching of silicon: A review. *Adv. Mater.* **2011**, *23*, 285–308. [CrossRef]
12. Lv, J.; Zhang, T.; Zhang, P.; Zhao, Y.; Li, S. Review application of nanostructured black silicon. *Nanoscale Res. Lett.* **2018**, *13*, 110. [CrossRef]
13. Jeong, Y.; Hong, C.; Jung, Y.H.; Akter, R.; Yoon, H.; Yoon, I. Enhanced Surface Properties of Light-Trapping Si Nanowires Using Synergetic Effects of Metal-Assisted and Anisotropic Chemical Etchings. *Sci. Rep.* **2019**, *9*, 15914. [CrossRef]
14. Hung, Y.J.; Lee, S.L.; Thibeault, B.J.; Coldren, L.A. Fabrication of highly ordered silicon nanowire arrays with controllable sidewall profiles for achieving low-surface reflection. *IEEE J. Sel. Top. Quantum Electron.* **2011**, *17*, 869–877. [CrossRef]
15. Smyrnakis, A.; Almpanis, E.; Constantoudis, V.; Papanikolaou, N.; Gogolides, E. Optical properties of high aspect ratio plasma etched silicon nanowires: Fabrication-induced variability dramatically reduces reflectance. *Nanotechnology* **2015**, *26*, 85301. [CrossRef]
16. Schmidt, V.; Wittemann, J.V.; Senz, S.; Gósele, U. Silicon nanowires: A review on aspects of their growth and their electrical properties. *Adv. Mater.* **2009**, *21*, 2681–2702. [CrossRef]
17. Bui, T.T.; Tu, H.P.; Dang, M.C. DRIE process optimization to fabricate vertical silicon nanowires using gold nanoparticles as masks. *Adv. Nat. Sci. Nanosci. Nanotechnol.* **2015**, *6*, 45016. [CrossRef]
18. Wang, D.-S.; Chao, J.-J.; Hung, S.-C.; Lin, C.-F. Fabrication of large-area gallium arsenide nanowires using silicon dioxide nanoparticle mask. *J. Vac. Sci. Technol. B Microelectron. Nanometer Struct. Process. Meas. Phenom.* **2009**, *27*, 2449. [CrossRef]
19. Baba, M.; Jia, T.; Suzuki, M.; Kuroda, H. Femtosecond Laser Induced Nanowire Technique and Its Applications. *ISRN Nanotechnol.* **2011**, *2011*, 907390. [CrossRef]
20. Toor, F.; Miller, J.B.; Davidson, L.M.; Duan, W.; Jura, M.P.; Yim, J.; Forziati, J.; Black, M.R. Metal assisted catalyzed etched (MACE) black Si: Optics and device physics. *Nanoscale* **2016**, *8*, 15448–15466. [CrossRef]
21. Duran, J.M.; Sarangan, A. Fabrication of ultrahigh aspect ratio silicon nanostructures using self-assembled gold metal-assisted chemical etching. *J. Micro/Nanolithogr. MEMS MOEMS* **2017**, *16*, 014502. [CrossRef]
22. Liu, X.; Coxon, P.R.; Peters, M.; Hoex, B.; Cole, J.M.; Fray, D.J. Black silicon: Fabrication methods, properties and solar energy applications. *Energy Environ. Sci.* **2014**, *7*, 3223–3263. [CrossRef]
23. Lin, P.; Xie, X.; Wang, Y.; Lian, B.; Zhang, G. A multi-step etch method for fabricating slightly tapered through-silicon vias based on modified Bosch process. *Microsyst. Technol.* **2019**, *25*, 2693–2698. [CrossRef]
24. Kim, K.; Lee, J.K.; Han, S.J.; Lee, S. A novel top-down fabrication process for vertically-stacked silicon-nanowire array. *Appl. Sci.* **2020**, *10*, 1146. [CrossRef]
25. Laermer, F.; Urban, A. MEMS at Bosch—Si plasma etch success story, history, applications, and products. *Plasma Process. Polym.* **2019**, *16*, 1–12. [CrossRef]
26. Garnett, E.; Yang, P. Light trapping in silicon nanowire solar cells. *Nano Lett.* **2010**, *10*, 1082–1087. [CrossRef]
27. Garnett, E.C.; Brongersma, M.L.; Cui, Y.; McGehee, M.D. Nanowire solar cells. *Annu. Rev. Mater. Res.* **2011**, *41*, 269–295. [CrossRef]
28. Flick, J.; Rivera, N.; Narang, P. Strong light-matter coupling in quantum chemistry and quantum photonics. *Nanophotonics* **2018**, *7*, 1479–1501. [CrossRef]

29. Zhang, Z.; Wang, Y.; Hansen, P.A.S.; Du, K.; Gustavsen, K.R.; Liu, G.; Karlsen, F.; Nilsen, O.; Xue, C.; Wang, K. Black silicon with order-disordered structures for enhanced light trapping and photothermic conversion. *Nano Energy* **2019**, *65*, 103992. [CrossRef]
30. Zhang, Z.; Martinsen, T.; Liu, G.; Tayyib, M.; Cui, D.; de Boer, M.J.; Karlsen, F.; Jakobsen, H.; Xue, C.; Wang, K. Ultralow Broadband Reflectivity in Black Silicon via Synergy between Hierarchical Texture and Specific-Size Au Nanoparticles. *Adv. Opt. Mater.* **2020**, *8*, 2000668. [CrossRef]
31. Furube, A.; Hashimoto, S. Insight into plasmonic hot-electron transfer and plasmon molecular drive: New dimensions in energy conversion and nanofabrication. *NPG Asia Mater.* **2017**, *9*, e454. [CrossRef]
32. Liu, H.; Zhang, X.; Zhai, T.; Sander, T.; Chen, L.; Klar, P.J. Centimeter-scale-homogeneous SERS substrates with seven-order global enhancement through thermally controlled plasmonic nanostructures. *Nanoscale* **2014**, *6*, 5099–5105. [CrossRef] [PubMed]
33. Yu, P.; Yao, Y.; Wu, J.; Niu, X.; Rogach, A.L.; Wang, Z. Effects of Plasmonic Metal Core-Dielectric Shell Nanoparticles on the Broadband Light Absorption Enhancement in Thin Film Solar Cells. *Sci. Rep.* **2017**, *7*, 7696. [CrossRef] [PubMed]
34. Schutzius, T.M.; Jung, S.; Maitra, T.; Graeber, G.; Köhme, M.; Poulikakos, D. Spontaneous droplet trampolining on rigid superhydrophobic surfaces. *Nature* **2015**, *527*, 82–85. [CrossRef]
35. Hao, C.; Li, J.; Liu, Y.; Zhou, X.; Liu, Y.; Liu, R.; Che, L.; Zhou, W.; Sun, D.; Li, L.; et al. Superhydrophobic-like tunable droplet bouncing on slippery liquid interfaces. *Nat. Commun.* **2015**, *6*, 7986. [CrossRef]
36. Wei, W.; Zhu, Y.; Wang, N.; Mei, H.; Yin, S. Photothermal characteristics of novel flexible black silicon for solar thermal receiver. *Int. J. Thermophys.* **2012**, *33*, 2179–2184. [CrossRef]
37. Ferraro, A.; Cerza, P.; Mussi, V.; Maiolo, L.; Convertino, A.; Caputo, R. Efficient Photothermal Generation by Nanoscale Light Trapping in a Forest of Silicon Nanowires. *J. Phys. Chem. C* **2021**, *125*, 25. [CrossRef]
38. Liu, D.; Li, L.; Gao, Y.; Wang, C.; Jiang, J.; Xiong, Y. The nature of photocatalytic "water splitting" on silicon nanowires. *Angew. Chem. Int. Ed.* **2015**, *54*, 2980–2985. [CrossRef]
39. Gaidi, M.; Daoudi, K.; Columbus, S.; Hajjaji, A.; Khakani, M.A.E.; Bessais, B. Enhanced photocatalytic activities of silicon nanowires/graphene oxide nanocomposite: Effect of etching parameters. *J. Environ. Sci.* **2021**, *101*, 123–134. [CrossRef]
40. Soam, A.; Arya, N.; Singh, A.; Dusane, R. Fabrication of silicon nanowires based on-chip micro-supercapacitor. *Chem. Phys. Lett.* **2017**, *678*, 46–50. [CrossRef]
41. Liu, R.; Wang, J.; Sun, T.; Wang, M.; Wu, C.; Zou, H.; Song, T.; Zhang, X.; Lee, S.T.; Wang, Z.L.; et al. Silicon Nanowire/Polymer Hybrid Solar Cell-Supercapacitor: A Self-Charging Power Unit with a Total Efficiency of 10.5%. *Nano Lett.* **2017**, *17*, 4240–4247. [CrossRef]
42. Lu, P.; Müller, L.; Hoffmann, M.; Chen, X. Taper silicon nano-scaffold regulated compact integration of 1D nanocarbons for improved on-chip supercapacitor. *Nano Energy* **2017**, *41*, 618–625. [CrossRef]
43. Li, L.; Tian, F.; Chang, H.; Zhang, J.; Wang, C.; Rao, W.; Hu, H. Interactions of Bacteria With Monolithic Lateral Silicon Nanospikes Inside a Microfluidic Channel. *Front. Chem.* **2019**, *7*, 483. [CrossRef]
44. Sun, N.; Lee, Y.T.; Zhang, R.Y.; Kao, R.; Teng, P.C.; Yang, Y.; Yang, P.; Wang, J.J.; Smalley, M.; Chen, P.J.; et al. Purification of HCC-specific extracellular vesicles on nanosubstrates for early HCC detection by digital scoring. *Nat. Commun.* **2020**, *11*, 4489. [CrossRef]
45. Zhu, K.; Zhang, Y.; Li, Z.; Zhou, F.; Feng, K.; Dou, H.; Wang, T. Simultaneous detection of α-fetoprotein and carcinoembryonic antigen based on Si nanowire field-effect transistors. *Sensors* **2015**, *15*, 19225–19236. [CrossRef]

linkcolor=black, citecolor=black, urlcolor=black,

micromachines

MDPI

Article

Deposition of Very-Low-Hydrogen-Containing Silicon at a Low Temperature Using Very-High-Frequency (162 MHz) SiH_4 Plasma

Ki Seok Kim, You-Jin Ji, Ki-Hyun Kim, Ji-Eun Kang, Albert Rogers Ellingboe and Geun Young Yeom

[1] Research Laboratory of Electronics, Massachusetts Institute of Technology, Cambridge, MA 02139, USA; kiseok@mit.edu
[2] School of Advanced Materials Science and Engineering, Sungkyunkwan University, 2066 Seobu-ro, Jangan-gu, Suwon-si 16419, Korea; gotnr3205@skku.edu (Y.-J.J.); ngdu@skku.edu (K.-H.K.); wldms8999@skku.edu (J.-E.K.)
[3] Plasma Research Laboratory, School of Physical Sciences and NCPST, Dublin City University, D09 V209 Dublin, Ireland; albert.ellingboe@dcu.ie
[4] SKKU Advanced Institute of Nano Technology (SAINT), Sungkyunkwan University, 2066 Seobu-ro, Jangan-gu, Suwon-si 16419, Korea
* Correspondence: gyyeom@skku.edu
† These authors contributed equally to this work.

Abstract: Low-hydrogen-containing amorphous silicon (a-Si) was deposited at a low temperature of 80 °C using a very high frequency (VHF at 162 MHz) plasma system with multi-split electrodes. Using the 162 MHz VHF plasma system, a high deposition rate of a-Si with a relatively high deposition uniformity of 6.7% could be obtained due to the formation of high-ion-density (>10^{11} cm^{-3}) plasma with SiH_4 and a lack of standing waves by using small multi-split electrodes. The increase in the radio frequency (RF) power decreased the hydrogen content in the deposited silicon film and, at a high RF power of 2000 W, a-Si with a low hydrogen content of 3.78% could be deposited without the need for a dehydrogenation process. The crystallization of the a-Si by ultraviolet (UV) irradiation showed that the a-Si can be crystallized with a crystallinity of 0.8 and a UV energy of 80 J without dehydrogenation. High-resolution transmission electron microscopy showed that the a-Si deposited by the VHF plasma was a very small nanocrystalline-like a-Si and the crystalline size significantly grew with the UV irradiation. We believe that the VHF (162 MHz) multi-split plasma system can be used for a low-cost low-temperature polysilicon (LTPS) process.

Citation: Ki Seok Kim, You-Jin Ji, Ki-Hyun Kim, Ji-Eun Kang, Albert Rogers Ellingboe and Geun Young Yeom Deposition of Very-Low-Hydrogen-Containing Silicon at a Low Temperature Using Very-High-Frequency (162 MHz) SiH_4 Plasma. *Micromachines* 2022, 13, 173. https://doi.org/10.3390/mi13020173

Academic Editor: Nikolaos Tapoglou

Received: 24 December 2021
Accepted: 20 January 2022
Published: 24 January 2022

Publisher's Note: MDPI stays neutral with regard to jurisdictional claims in published maps and institutional affiliations.

Keywords: very high frequency (VHF); multi-split electrode; low-temperature polysilicon (LTPS)

1. Introduction

Among thin-film transistors (TFTs), low-temperature polysilicon (LTPS) TFTs have significant advantages in device characteristics, such as a low threshold voltage, steep subthreshold swing, high voltage/current reliability, and high yield [1]. They are widely applied in high-resolution liquid crystal displays (LCDs), image sensors, photovoltaics, active-matrix organic light emitting diodes (AMOLEDs), and to flexible displays, because their mobility is about 100 times faster than that of a-Si thin-film transistors (TFTs) [2,3].

The process steps used for fabricating LTPS TFTs consist of (1st step) a-Si deposition; (2nd step) dehydrogenation; and (3rd step) crystallization using various methods, such as rapid thermal annealing (RTA) [4], metal-induced lateral crystallization (MILC) [5,6], plasma surface treatment [7,8], and lamp heating [9–12], as well as excimer laser annealing (ELA) [2,13–15]. Among these, a-Si is generally deposited by plasma-enhanced chemical vapor deposition (PECVD). The a-Si deposited by PECVD generally contains hydrogen at ~10% or more [2]. When the crystallization process is carried out using high hydrogen containing a-Si for polysilicon formation, hydrogen diffuses quickly, which makes the surface rough and porous [16]. Therefore, a dehydrogenation process is required and the process is performed during a lengthy heat treatment at ~450 °C before the crystallization process, which results in high equipment investment costs.

In this study, for the simplification of the LTPS process requiring no dehydration, a-Si was deposited at a low temperature of 80 °C using a very high frequency (VHF, 162 MHz) plasma system with multi-split electrodes. By using a VHF (162 MHz) plasma system, it is possible to form a high-quality thin film by increasing the gas dissociation rate compared to that of conventional 13.56 MHz plasma. The advantages of the use of VHF plasma compared to conventional 13.56 MHz in thin-film deposition and surface treatment have been clearly demonstrated by our previous studies [17–19], and the reasons for this are the low electron temperature and high vibrational temperature of VHF (162 MHz) plasma [19]. Since the VHF plasma generated with multi-split electrodes can sufficiently decompose the silane (SiH_4) gas without showing a standing wave effect [20–22] and, therefore, can deposit a-Si with a very low hydrogen content uniformly, highly crystallized polysilicon could be fabricated through a crystallization process using an ultraviolet (UV) lamp without a dehydrogenation process.

2. Materials and Methods

Figure 1 is a schematic diagram of the VHF (162 MHz) plasma system with multi-split electrodes used in this experiment. The 162 MHz RF power was equally supplied to each multi-split electrode through the matcher and a power splitter system. A detailed description of the VHF (162 MHz) plasma system with multi-split electrodes can be found elsewhere [15,16,22–25]. Each electrode was made of a rectangular tile-shaped (11 cm × 12 cm) anodized aluminum with holes for the showerhead to distribute SiH_4 gas uniformly in the process chamber. The electrodes were located ~1 cm apart regularly in pairs, and the split RF power was applied to two adjacent tile pair electrodes (shown as blue and white tiles in Figure 2).

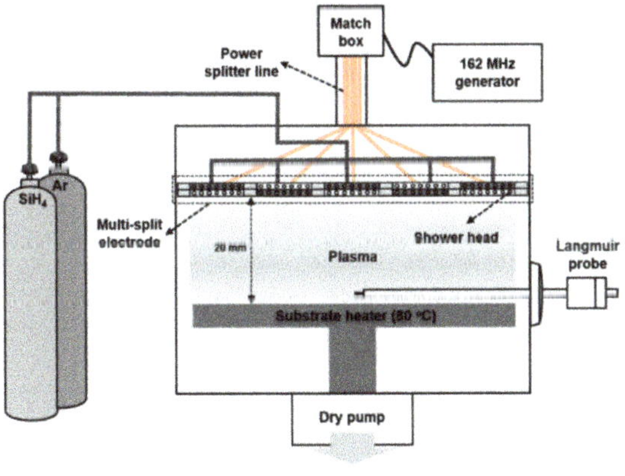

Figure 1. Schematic diagram of the VHF (162 MHz) plasma system with multi-split electrodes.

The a-Si was deposited on Corning glass substrates at the operating pressure of 200 mTorr SiH_4 and a low temperature of 80 °C for 2 min while the 162 MHz RF power was varied from 500 to 2000 W. The deposited a-Si films were crystallized by ultraviolet (UV) irradiation using a UV lamp with a wavelength of 254 nm. The energy of the UV lamp was varied from 50 to 80 J.

The surface morphology of the deposited a-Si was observed by field emission scanning electron microscopy (FE-SEM, S4700, Hitachi, Tokyo, Japan). Secondary ion mass spectroscopy (D-SIMS, IMS 4FE7, Cameca, Gennevilliers, France) was carried out to extract the hydrogen contents in the deposited a-Si films using Cs^+ ions with an impact energy of 14.5 kV. X-ray photoelectron spectroscopy (XPS, MultiLab 2000, Thermo VG, Waltham,

MA USA, Mg Kα source) was performed to evaluate the chemical properties of deposited a-Si. The crystallinity of the UV-irradiated a-Si was measured by Raman spectroscopy (α 300 M +, WITEC, Eglfinger weg2, Germany) and transmittance electron microscopy (TEM, JEM-2100F, JEOL, Tokyo, Japan). A Langmuir probe (ALP-150, Impedans, Dublin, Ireland) was used to measure the ion density in the SiH_4 plasmas.

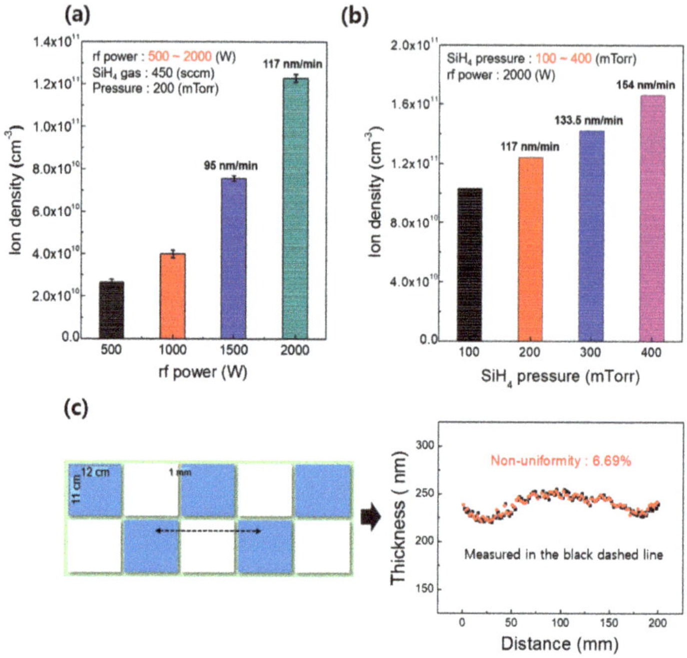

Figure 2. Ion density measured using a Langmuir probe (**a**) at different RF powers (500~2000 W) and (**b**) different SiH_4 pressures (100~400 mTorr) in a VHF (162 MHz) plasma source with multi-split electrodes. (**c**) The schematic diagram of multi-split electrodes, with the thickness/non-uniformity measured at 1 mm intervals in a 200 mm region of an a-Si thin film deposited using the VHF (162 MHz) plasma source at 2000 W, 200 mTorr, and 80 °C substrate temperature.

3. Results and Discussion

Figure 2a,b show the ion density of SiH_4 plasmas measured as a function of RF power at a fixed SiH_4 pressure of 200 mTorr and as a function of SiH_4 pressure at a fixed RF power of 2000 W, respectively. The ion density was measured using a Langmuir probe at the center of the process chamber near the substrate location. As shown in Figure 2a, the ion density was gradually increased by increasing the RF power from 500 to 2000, and a high plasma density of $\geq 10^{11}$ cm^{-3} could be obtained at 2000 W. The growth rates of a-Si at 1500 and 2000 W with a fixed SiH_4 pressure of 200 mTorr were measured as 95 and 117 nm/min, respectively. Additionally, as shown in Figure 2b, as the SiH_4 pressure was increased from 100 to 400 mTorr at a fixed RF power of 2000 W, the ion density was increased and a high ion density (> 10^{11} cm^{-3}) was obtained at a pressure over 200 mTorr. On the other hand, conventional 13.56 MHz CCP showed a low plasma density at around 10^{10} cm^{-3} [26]. As the pressure increased from 200 to 400 mTorr, the growth rate also increased from 117 to 154 nm/min.

Figure 2c shows the thickness uniformity of a-Si film deposited at a fixed condition of 2000 W of RF power and 200 mTorr of SiH_4. The deposition process was performed at the substrate temperature of 80 °C for 2 min using the rectangular multi-split electrodes shown in Figure 2c. The thickness of the deposited a-Si was measured along the black dash line (200 mm) with 1 mm intervals to evaluate the thickness uniformity across the electrode. The points of measurement include the positions of repeating electrode tile faces/boundaries. As shown in Figure 2c, the measured thickness non-uniformity was 6.69% and the average thickness of a-Si was 234 nm. In general, when a VHF is used, due to the short wavelength of the VHF compared to the size of the electrode, a standing wave effect causing non-uniformity along the electrode centerline can be observed. By using the multi-split electrodes with a size much smaller than the wavelength of the VHF, the standing wave effect was suppressed and resulting film was relatively uniform. Moreover, a high deposition rate of 117 nm/min was achieved due to the high dissociation rate of SiH_4, even at a low deposition temperature. In our previous studies, we have demonstrated that 162 MHz plasma has significantly higher dissociation rates than 60 and 13.56 MHz plasmas, which can minimize hydrogen bonding in the thin film while improving the deposition rate [15,16].

To investigate the influence of RF power on the hydrogen contents of a-Si, a-Si was deposited with different RF powers of 1000~2000 W. Figure 3a,b show SIMS data (left) on H, O, and Si for a-Si deposited with the RF powers of 1500 and 2000 W and the hydrogen concentration in the deposited a-Si calculated from the SIMS results (right). Using the relation of $C_H = (N_H/N_{si}) \times 100$, where N_H is from SIMS results in Figure 3b and N_{Si} is $5 \times 10^{22}/cm^3$ from the literature [27,28], the hydrogen contents in the deposited a-Si films were calculated and the results are shown in Figure 3c. As shown in Figure 3c, the calculated hydrogen content in the a-Si film was decreased from 5.3 to 3.78% with the increasing RF power, which shows that the RF power can influence the hydrogen content in the films. During the deposition process, hydrogen atoms weakly bonded with Si atoms can be easily eliminated by ion bombardment. High RF power not only increases the dissociation rate of SiH_4 but also increases the ion densities as shown in Figure 2a, causing more ion bombardment on the depositing a-Si surface in addition to increased atomic hydrogen flux [27–30]. Therefore, at the higher RF power conditions, H atoms adsorbed on the a-Si surface could be easily desorbed from a-Si surface, because the rate of H desorption is influenced by ion bombardment [31–33]. Figure 3d–f show the XPS narrow scan data of Si 2p of a-Si deposited with RF powers of 1000~2000 W. The XPS narrow scan data of the Si 2p were deconvoluted into Si 2p and Si compounds, which are composed of Si-C, Si-O, and Si-OH near the binding energy of 101~104 eV [34]. In the VHF (162 MHz) multi-split plasma system, the increase in RF power increased the dissociation rate of the SiH_4 and decreased the content of Si compounds, indicating the growth of high-quality a-Si films at higher RF powers. Figure 3g,h show the top-view SEM images of a-Si corresponding to Figure 3a,b), respectively. Clearly, more nanocrystalline-like a-Si was observed at a higher RF power.

Based on the results obtained for the a-Si with a lower hydrogen content deposited at the higher RF power, the UV crystallization process was performed on the deposited a-Si film under the condition of an RF power of 2000 W, SiH_4 flow rate of 450 sccm, working pressure of 200 mTorr, and substrate temperature of 80 °C. For the crystallization of the low-hydrogen-containing a-Si, a UV lamp with a wavelength of 254 nm was used to crystallize a-Si and the UV lamp was exposed for 1 min at a distance of 1 cm from the a-Si. Figure 4a shows the Raman spectra of the a-Si films before and after irradiation at different UV lamp energies to confirm the crystalline property of the irradiated a-Si. Untreated pristine a-Si shows a broad and low intensity of peaks centered at 492.78 cm^{-1}. In contrast, the a-Si irradiated by the UV lamp with the energy of 50~80 J showed a relatively sharp peak blue-shifted with respect to the pristine a-Si, and the peak positions were closer to the crystalline Si Raman peak of 520 cm^{-1} at increasing UV energies. At the UV energy of 80 J, a sharp peak at 516.81 cm^{-1} was observed, indicating the high crystallinity of UV irradiated a-Si. Raman crystallinity can be estimated from the Raman shift of the UV irradiated a-Si by applying Equation (1) [35–38]:

$$X_c = (I_{500} + I_{520})/(aI_{480} + I_{500} + I_{520}) \quad (1)$$

where I_{480}, I_{500}, and I_{520} correspond to the amorphous, intermediate, and crystalline components, respectively. From the above equation, the crystallinities of a-Si irradiated with different UV energies are plotted in Figure 4b. As shown in Figure 4b, the Raman crystallinity was gradually improved from 0.45 to 0.8 as the UV energy was increased from 50 to 80 J.

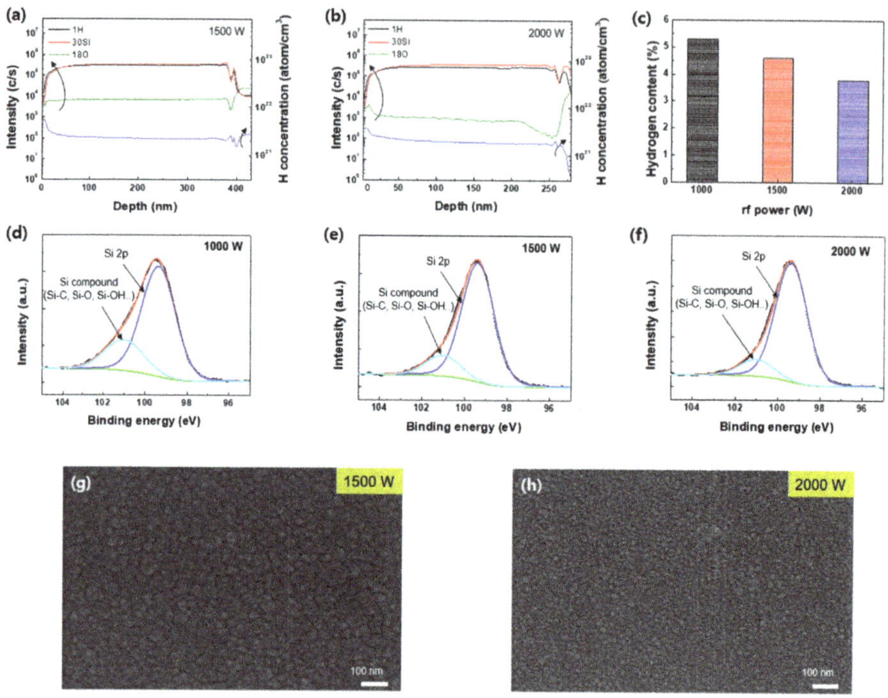

Figure 3. SIMS depth profile of a-Si films with different RF powers of (**a**) 1500 W and (**b**) 2000 W. (**c**) Hydrogen contents and (**d–f**) XPS Si 2p spectra of deposited a-Si at different RF powers (1000~2000 W). SEM images of a-Si deposited at (**g**) 1500 W and (**h**) 2000 W.

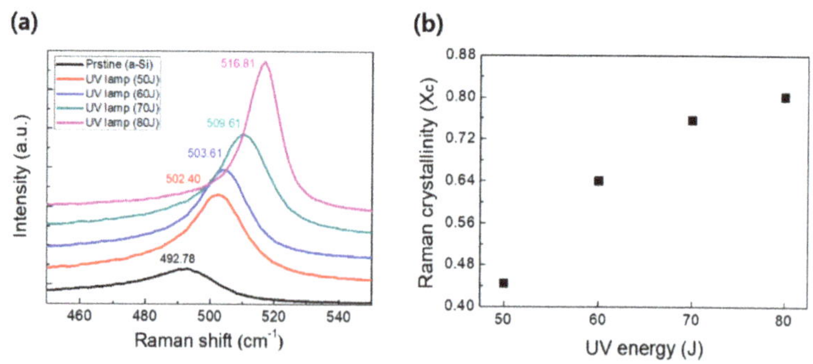

Figure 4. (**a**) Raman spectra and (**b**) Raman crystallinity (X_c) of a-Si after the crystallization with different UV energies (50~80 J).

An HR-TEM analysis was performed to confirm the morphology of the silicon thin films crystallized after the irradiation using the UV lamp. Figure 5 shows the HR-TEM images and FFT patterns of the a-Si films before and after the crystallization by the UV lamp (60 and 80 J). Figure 5a shows the small nanocrystalline-like grains of the a-Si before the UV exposure deposited using 2000 W of VHF (162 MHz) SiH_4 plasma. In order to more easily distinguish the grain size, noticeable grain boundaries are shown as a red line in Figure 5d. Figure 5g is the FFT pattern measured for Figure 5d, and an amorphous ring pattern was confirmed due to the nanocrystalline-like grains. On the other hand, as shown in the HR-TEM images of Figure 5e,f, the grain size was gradually increased as the UV lamp energy was increased to 60 and 80 J, respectively, and strong polycrystalline phases were also observed in the FFT patterns, as shown in Figure 5h,i. These results demonstrate that low-hydrogen-containing a-Si deposited using a VHF (162 MHz) multi-split plasma source can be crystallized without a dehydrogenation process.

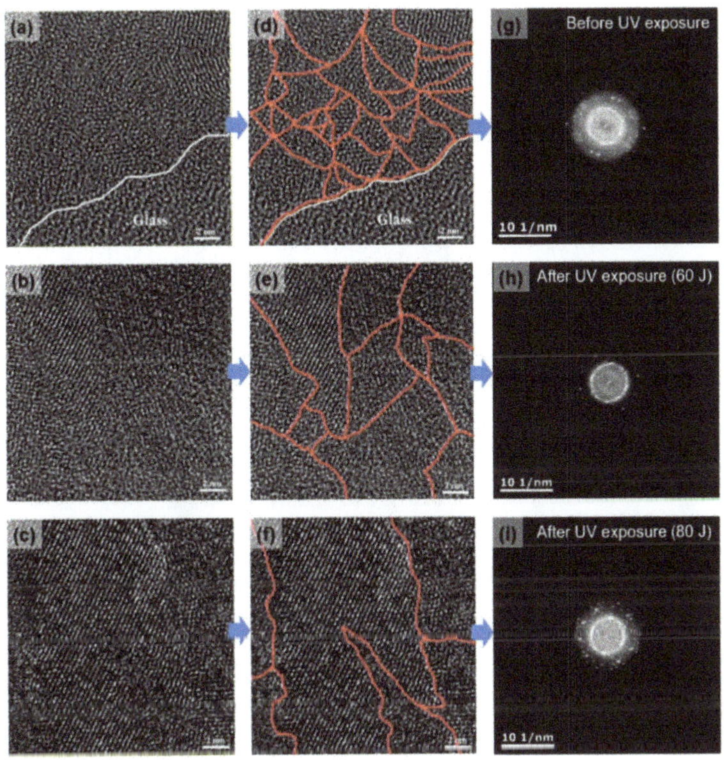

Figure 5. (**a–c**) HR-TEM image, (**d–f**) grain boundary indicated (red line) in the HR-TEM image, and (**g–i**) fast Fourier transform (FFT) pattern of as-deposited a-Si irradiated by UV lamp at 60 and 80 J.

4. Conclusions

VHF (162 MHz) SiH_4 plasmas with multi-split electrodes were investigated regarding the deposition of low-hydrogen-containing a-Si required for the crystallization of silicon without the use of a dehydrogenation process at a low substrate temperature of 80 °C uniformly without showing the standing wave effect. Increasing the VHF RF power to the SiH_4 plasma not only increased the hydrogen dissociation but also increased the ion density in the plasma. By increasing the RF power in the deposited a-Si film, the hydrogen percentage was decreased to 3.78% at 2000 W of RF power, while a decrease in the impurities in the film, such as carbon/oxygen, was also achieved. The low content of impurities such as hydrogen, carbon, and oxygen in the a-Si deposited at the higher RF power was related to the high ion density ($>10^{11}$ cm^{-3}), which led to the removal of the hydrogen adsorbed during the growth of a-Si in addition to the removal of other impurities such as carbon/oxygen by the enhanced ion bombardment of the substrate. When the a-Si was crystallized with a UV lamp with the low-hydrogen-containing a-Si, the improved crystallinity was confirmed by Raman spectroscopy and HR-TEM, and the crystallinity of 0.8 could be obtained by Raman crystallinity.

Author Contributions: Conceptualization, K.S.K. and G.Y.Y.; methodology, K.S.K. and Y.-J.J.; investigation, J.-E.K. and K.-H.K.; data curation, K.S.K. and Y.-J.J.; writing-original draft, K.S.K. and Y.-J.J.; writing—review and editing, K.S.K., Y.-J.J., K.-H.K. and G.Y.Y.; supervision, A.R.E. and G.Y.Y. All authors have read and agreed to the published version of the manuscript.

Funding: This work was supported by the Technology Innovation Program (20014639, Development of high density inorganic thin film deposition system without micro defects) funded by the Ministry of Trade, Industry & Energy (MOTIE, Korea).

Data Availability Statement: The data presented in this study are available on request from the corresponding author upon reasonable request.

Conflicts of Interest: The authors declare no conflict of interest.

References

1. Kim, H.; Jeong, D.Y.; Lee, S.; Jang, J. A High-Gain inverter with low-temperature poly-Si oxide thin-film transistors. *IEEE Electron Device Lett.* **2019**, *40*, 411–414. [CrossRef]
2. Yogoro, Y.; Masuda, A.; Matsumura, H. Crystallization by excimer laser annealing for a-Si: H films with low hydrogen content prepared by Cat-CVD. *Thin Solid Films* **2003**, *430*, 296–299. [CrossRef]
3. Brotherton, S.D. Polycrystalline silicon thin film transistors. *Semicond. Sci. Technol.* **1995**, *10*, 721. [CrossRef]
4. Liu, G.; Fonash, S.J. Selective area crystallization of amorphous silicon films by low-temperature rapid thermal annealing. *Appl. Phys. Lett.* **1989**, *55*, 660–662. [CrossRef]
5. Nast, O.; Wenham, S.R. Elucidation of the layer exchange mechanism in the formation of polycrystalline silicon by aluminum-induced crystallization. *J. Appl. Phys.* **2000**, *88*, 124–132. [CrossRef]
6. Sun, K.; Hakim, M.M.A.; Gunn, R.; Ashburn, P. Effect of an oxide cap layer and fluorine implantation on the metal-induced lateral crystallization of amorphous silicon. *ECS J. Solid State Sci. Technol.* **2013**, *2*, P42–P45. [CrossRef]
7. Lin, C.W.; Yang, M.Z.; Yeh, C.C.; Cheng, L.J.; Huang, T.Y.; Cheng, H.C.; Lin, H.C.; Chao, T.S.; Chang, C.Y. Effects of plasma treatments, substrate types, and crystallization methods on performance and reliability of low temperature polysilicon TFTs. In Proceedings of the International Electron Devices Meeting, Washington, DC, USA, 5–8 December 1999.
8. Nakashima, R.; Shin, R.; Hanafusa, H.; Higashi, S. Generation of ultra high-power thermal plasma jet and its application to crystallization of amorphous silicon films. *Jpn. J. Appl. Phys.* **2017**, *56*, 06HE05. [CrossRef]
9. Hwang, J.H.; Kim, H.J.; Kim, B.K.; Jin, W.B.; Kim, Y.; Chung, H.; Park, S. Scanning multishot irradiations on a large-area glass substrate for Xe-Arc flash lamp crystallization of amorphous silicon thin-film. *Int. J. Therm. Sci.* **2015**, *91*, 1–11. [CrossRef]
10. Pécz, B.; Dobos, L.; Panknin, D.; Skorupa, W.; Lioutas, C.; Vouroutzis, N. Crystallization of amorphous-Si films by flash lamp annealing. *Appl. Surf. Sci.* **2005**, *242*, 185–191. [CrossRef]
11. Shim, M.S.; Kim, C.H.; Yi, J. UV crystallization of poly-si using a CeO_2 seed layer on plastic substrate for microelectronics applications. *Thin Solid Films* **2004**, *453*, 100–105.
12. Khakifirooz, A.; Haji, S.; Mohajerzadeh, S.S. UV-assisted nickel-induced crystallization of amorphous silicon. *Thin Solid Films* **2001**, *383*, 241–243. [CrossRef]
13. Park, K.C.; Lee, J.H.; Song, I.H.; Jung, S.H.; Han, M.K. Poly-Si thin film transistors fabricated by combining excimer laser annealing and metal induced lateral crystallization. *J. Non-Cryst. Solids* **2002**, *299*, 1330–1334. [CrossRef]
14. Moschou, D.C.; Exarchos, M.A.; Kouvatsos, D.N.; Papaioannou, G.J.; Voutsas, A.T. Performance and reliability of SLS ELA polysilicon TFTs fabricated with novel crystallization techniques. *Microelectron. Reliab.* **2007**, *47*, 1378–1383. [CrossRef]
15. Exarchos, M.A.; Moschou, D.C.; Papaioannou, G.J.; Kouvatsos, D.N.; Arapoyanni, A.; Voutsas, A.T. On the study of p-channel thin-film transistors fabricated by SLS ELA crystallization techniques. *Thin Solid Films* **2009**, *517*, 6375–6378. [CrossRef]
16. Lengsfeld, P.; Nickel, N.H.; Fuhs, W. Step-by-step excimer laser induced crystallization of a-Si:H. *Appl. Phys. Lett.* **2000**, *76*, 1680–1682. [CrossRef]
17. Ji, Y.J.; Kim, K.S.; Kim, K.H.; Ellingboe, A.R.; Yeom, G.Y. Nitriding process for next-generation semiconductor devices by VHF (162 MHz) multi-tile push-pull plasma source. *Appl. Surf. Sci.* **2020**, *506*, 144904. [CrossRef]
18. Kim, K.S.; Sirse, N.; Kim, K.H.; Ellingboe, A.R.; Kim, K.N.; Yeom, G.Y. Characteristics of silicon nitride deposited by VHF (162 MHz)-plasma enhanced chemical vapor deposition using a multi-tile push–pull plasma source. *J. Phys. D Appl. Phys.* **2016**, *49*, 395201. [CrossRef]
19. Byun, J.Y.; Ji, Y.J.; Kim, K.H.; Kim, K.S.; Tak, H.W.; Ellingboe, A.R.; Yeom, G.Y. Characteristics of silicon nitride deposited by very high frequency (162 MHz)-plasma enhanced atomic layer deposition using bis (diethylamino) silane. *Nanotechnology* **2020**, *32*, 075706. [CrossRef]
20. Lieberman, M.A.; Booth, J.P.; Chabert, P.; Rax, J.M.; Turner, M.M. Standing wave and skin effects in large-area, high-frequency capacitive discharges. *Plasma Sources Sci. Technol.* **2002**, *11*, 283. [CrossRef]

21. Sansonnens, L.; Schmidt, H.; Howling, A.A.; Hollenstein, C.; Ellert, C.; Buechel, A. Application of the shaped electrode technique to a large area rectangular capacitively coupled plasma reactor to suppress standing wave nonuniformity. *J. Vac. Sci. Technol. A* **2006**, *24*, 1425–1430. [CrossRef]
22. Kim, K.S.; Kim, K.H.; Ji, Y.J.; Park, J.W.; Shin, J.H.; Ellingboe, A.R.; Yeom, G.Y. Silicon nitride deposition for flexible organic electronic devices by VHF (162 MHz)-PECVD using a multi-tile push-pull plasma source. *Sci. Rep.* **2017**, *7*, 13585. [CrossRef]
23. Sirse, N.; Harvey, C.; Gaman, C.; Ellingboe, A.R. Investigation of plasma uniformity, rotational and vibrational temperature in a 162 MHz multi-electrode capacitive discharge. *J. Phys. D Appl. Phys.* **2020**, *53*, 335203. [CrossRef]
24. Harvey, C.; Sirse, N.; Gaman, C.; Ellingboe, A.R. Mode transition in an oxygen low-pressure, very high frequency (162 MHz), multi-tile electrode capacitively coupled plasma. *Phys. Plasma* **2020**, *27*, 110701. [CrossRef]
25. Harvey, C.; Sirse, N.; Gaman, C.; Ellingboe, A.R. Experimental study of a high-VHF multi-electrode capacitively coupled plasma for thin layer surface processing. In Proceedings of the 44th EPS Conference on Plasma Physics, Belfast, Northern Ireland, 26 June 2017.
26. Popov, O.A. *High Density Plasma Sources: Design, Physics and Performance*; Elsevier: Woburn, MA, USA, 1996.
27. Rüther, R.; Livingstone, J. Hydrogenated amorphous silicon: Hydrogen content, bonding configurations and morphology in sputter-deposited, in-chamber annealed thin films. *Thin Solid Films* **1994**, *251*, 30–35. [CrossRef]
28. Goh, B.T.; Wah, C.K.; Aspanut, Z.; Rahman, S.A. Structural and optical properties of nc-Si: H thin films deposited by layer-by-layer technique. *J. Mater. Sci. Mater. Electron.* **2014**, *25*, 286–296. [CrossRef]
29. Chen, C.Z.; Qiu, S.H.; Liu, C.Q.; Wu, Y.D.; Li, P.; Yu, C.Y.; Lin, X.Y. Low temperature fast growth of nanocrystalline silicon films by rf-PECVD from SiH_4/H_2 gases: Microstructural characterization. *J. Phys. D Appl. Phys.* **2008**, *41*, 195413. [CrossRef]
30. Xuan-Ying, L.; Chuang-Jun, H.; Kui-Xun, L.; Yun-Peng, Y.; Chu-Ying, Y.; Ling-Fei, C. Low-temperature growth of polycrystalline silicon films by $SiCl_4/H_2$ rf plasma enhanced chemical vapour deposition. *Chin. Phys. Lett.* **2003**, *20*, 1879. [CrossRef]
31. Andújar, J.L.; Bertran, E.; Canillas, A.; Roch, C.; Morenza, J.L. Influence of pressure and radio frequency power on deposition rate and structural properties of hydrogenated amorphous silicon thin films prepared by plasma deposition. *J. Vac. Sci. Technol. A* **1991**, *9*, 2216–2221. [CrossRef]
32. Aida, M.S.; Mahdjoubi, L.; Sahli, S. The effect of RF power on the hydrogen content of sputtered amorphous silicon. *Mater. Chem. Phys.* **1992**, *32*, 349–351. [CrossRef]
33. Hsiao, W.C.; Liu, C.P.; Wang, Y.L. Influence of RF bias on hydrogenated amorphous silicon by high-density plasma chemical vapor deposition. *J. Electrochem. Soc.* **2007**, *154*, G122. [CrossRef]
34. Chastain, J.; King, R.C., Jr. *Handbook of X-ray Photoelectron Spectroscopy*; Perkin-Elmer: Eden Prairie, MN, USA, 1992; 261p.
35. Monaghan, E.; Yeom, G.Y.; Ellingboe, A.R. Measurement of nc-Si: H film uniformity and diagnosis of plasma spatial structure produced by a very high frequency, differentially powered, multi-tile plasma source. *Vacuum* **2015**, *119*, 34–46. [CrossRef]
36. Wei, D.Y.; Xiao, S.Q.; Huang, S.Y.; Chan, C.S.; Zhou, H.P.; Xu, L.X.; Guo, Y.N.; Chai, J.W.; Wang, S.J.; Xu, S. Low-temperature deposition of μc-Si: H thin films by a low-frequency inductively coupled plasma for photovoltaic applications. *J. Phys. D Appl. Phys.* **2013**, *46*, 215501. [CrossRef]
37. Tong, G.B.; Rahman, S.A. Effects of RF power on the hydrogen bonding and structural order in Si: H thin films deposited from hydrogen diluted silane. In *AIP Conference Proceedings*; American Institute of Physics: College Park, MD, USA, 2010; Volume 1217, pp. 159–165.
38. Mukhopadhyay, S.; Chowdhury, A.; Ray, S. Substrate temperature dependence of microcrystalline silicon growth by PECVD technique. *J. Non-Cryst. Solids* **2006**, *352*, 1045–1048. [CrossRef]

micromachines

MDPI

Article

Fabrication of Three-Dimensionally Deformable Metal Structures Using Precision Electroforming

Seitaro Kumamoto [1,2], Souichiro Fukuyama [1], Seiya Nagano [1], Keiichiro Yasuda [2], Yusuke Kitamura [3], Masaaki Iwatsuki [4], Hideo Baba [4], Toshihiro Ihara [3], Yoshitaka Nakanishi [3,5] and Yuta Nakashima [3,5,6,7,*]

1 Graduate School of Science and Technology, Kumamoto University, 2-39-1 Kurokami, Chuo-ku, Kumamoto 860-8555, Japan; s.kumamoto@ogic.ne.jp (S.K.); souichirof1@yahoo.co.jp (S.F.); seiya-nagano@tamadic.co.jp (S.N.)
2 Ogic Technologies Co., Ltd., 2-9-9 Kamikumamoto, Nishi-ku, Kumamoto 860-0079, Japan; kyasuda@ogic.ne.jp
3 Faculty of Advanced Science and Technology, Kumamoto University, 2-39-1 Kurokami, Chuo-ku, Kumamoto 860-8555, Japan; ykita@kumamoto-u.ac.jp (Y.K.); toshi@chem.kumamoto-u.ac.jp (T.I.); y-naka@mech.kumamoto-u.ac.jp (Y.N.)
4 Faculty of Life Science, Kumamoto University, 1-1-1 Honjo, Chuo-ku, Kumamoto 860-8556, Japan; maiwa217@kumamoto-u.ac.jp (M.I.); hdobaba@kumamoto-u.ac.jp (H.B.)
5 Institute of Industrial Nanomaterials, Kumamoto University, 2-39-1 Kurokami, Chuo-ku, Kumamoto 860-8555, Japan
6 International Research Organization for Advanced Science & Technology, Kumamoto University, Kumamoto 860-8555, Japan
7 Fusion Oriented Research for Disruptive Science and Technology Researcher (FOREST Researcher), Japan Science and Technology Agency, Saitama 332-0012, Japan
* Correspondence: yuta-n@mech.kumamoto-u.ac.jp; Tel./Fax: +81-96-342-3743

Abstract: It is difficult to fabricate three-dimensional structures using semiconductor-process technology, because it is based on two-dimensional layered structure fabrication and the etching of thin films. In this study, we fabricated metal structures that can be dynamically deformed from two-dimensional to three-dimensional shapes by combining patterning using photolithography with electroforming technology. First, a resist structure was formed on a Cu substrate. Then, using a Ni sulfamate electroforming bath, a Ni structure was formed by electroforming the fabricated resist structure. Finally, the resist structure was removed to release the Ni structure fabricated on the substrate, and electroforming was used to Au-plate the entire surface. Scanning-electron microscopy revealed that the structure presented a high aspect ratio (thickness/resist width = 3.5), and metal structures could be fabricated without defects across the entire surface, including a high aspect ratio. The metallic structures had an average film thickness of 12.9 μm with $\sigma = 0.49$ μm, hardness of 600 HV, and slit width of 7.9 μm with $\sigma = 0.25$ μm. This microfabrication enables the fabrication of metal structures that deform dynamically in response to hydrodynamic forces in liquid and can be applied to fields such as environmental science, agriculture, and medicine.

Keywords: precision processing; electroforming; MEMS; metal structures; deformation

Citation: Kumamoto, S.; Fukuyama, S.; Nagano, S.; Yasuda, K.; Kitamura, Y.; Iwatsuki, M.; Baba, H.; Ihara, T.; Nakanishi, Y.; Nakashima, Y. Fabrication of Three-Dimensionally Deformable Metal Structures Using Precision Electroforming. *Micromachines* **2022**, *13*, 1046. https://doi.org/10.3390/mi13071046

Academic Editor: Nikolaos Tapoglou

Received: 25 May 2022
Accepted: 27 June 2022
Published: 30 June 2022

Publisher's Note: MDPI stays neutral with regard to jurisdictional claims in published maps and institutional affiliations.

1. Introduction

In fields such as electronic-equipment manufacturing, biotechnology, and medicine, miniaturization and large-scale integration of systems are advancing, and process technology is being used to upgrade the functions of materials, substrates, and microdevices [1–6]. This development includes the demand for various structures with complex shapes, such as three-dimensional structures for realizing effective microregional chemical reactions, mixing, and analyses in various fields, including biological microelectromechanical systems (bioMEMSs) and micro total analysis systems (μTASs) [7–11]. Studies on the fabrication of simple 2.5-dimensional structures with metals have been conducted, by devising methods for laminating and tapering to achieve these structures. However, these metal structures are static. Dynamic devices include structures in which polymers such as polydimethylsiloxane deform upon the application of pressure [12] and movable structures such as propellers that employ photopolymerization reactions [13]. However, although resin materials deform dynamically, it is impossible to perform various chemical modifications on structural surfaces. Single-crystal silicon can be combined with multiple materials to form a three-dimensional structure, but it is brittle. Although three-dimensionally deformable materials include comb actuators used in silicon material, chemical modification of the surfaces cannot be performed. In contrast, precision processing, such as cutting and grinding, may be used to produce free three-dimensional shapes for metal materials. However, such processing can cause variations by as much as 10 μm [14–16], which is unsuitable when the required precision is on the order of microns. One option for processing that can be realized with high precision and a high aspect ratio is deep reactive ion etching (DRIE) of silicon [17–20], used in anisotropic processing. DRIE is widely employed for producing dynamic random-access memories (DRAMs) and other devices. However, the cost of fabrication is a problem, as expensive equipment and time are required to achieve the necessary degree of vacuum. In comparison, high aspect ratios can be fabricated easily, as the combined photolithographic patterning and electroforming technique does not require a high vacuum or expensive equipment. However, the use of electroforming technology requires a seed-layer-forming process, which poses problems for productivity. The Lithographie, Galvanoformung, and Abformung (LIGA) process [21,22] is utilized when microelectromechanical system (MEMS)-level microfabrication techniques are required. However, X-ray equipment is expensive, posing problems for productivity and simplicity [23]. Precision electroforming technology, which combines photolithography and electroforming technology, has the advantages of better dimensional accuracy and higher productivity, compared with machining approaches such as cutting and grinding, and enables high-speed replication of the parent mold [24]. However, most extant studies on robust structures for miniaturization and precision improvement are concerned with fabricated structures [25–28], whereas those on structures with three-dimensional deformability are lacking. The microstructure accuracy has been reported to be limited to a 10% variation in a 50 μm design [29]. If a method with a higher aspect ratio and accuracy could be developed, microfabrication technology could be improved, and precision components could be produced. There are some examples of using precision electroforming technology to fabricate electric-shaver blades, but the structures are robust, require a hand-pushing force, and do not exhibit chemical modification. If a structure that deforms from a two-dimensional shape into a three-dimensional shape due to the hydrodynamic force could be produced via precision electroforming, the scope of the microdevice application (adaptation) would be expanded in a useful manner. μTAS has the advantage of reducing the amount of liquid required and shortening the analysis time by using a small amount of liquid. In microfluidic systems, the flow rate is approximately 1 $\mu L/min^{-1}$ mL/min [30–32]. The device must be agitated because it is difficult to mix the liquid when the width and depth of the flow channel decrease, the Reynolds number decreases, and the flow becomes laminar. For example, it would be advantageous to create a structure that dynamically agitates the inside of a microchannel to enable effective mixing. Furthermore, if it can dynamically deform into a three-dimensional structure or agitate, it can be applied to cases in which the liquid is fed in fixed quantities or filtered according to the flow rate. Therefore, in this study, we focused on the use of precision electroforming technology to investigate a method of fabricating an original metal structure capable of three-dimensional elastic deformation in response to a hydrodynamic force. Although flat-shaped structures can be fabricated using semiconductor processes, this method can produce structures that deform three-dimensionally by a hydrodynamic force. Furthermore, we evaluated its physical properties and basic operations.

2. Experimental Method

2.1. Design of a Three-Dimensional Elastically Deformable Metal Structure

Figure 1 presents the designs of the eight types of fabricated metallic structures. Types A and E consist of multiple slits arranged in comb-like shapes. The total length of the slits is approximately 6 mm, and the slit widths in Types A and E are 10 μm and 20 μm, respectively. In these structures, the root portion of the comb teeth is deformed at its origin by hydrodynamic forces, and plate-like structures mutually intersect. Types B and C are structures in which multiple slits are arranged in an arc shape, the slit width is 10 μm, and the pitches are 250 μm and 500 μm, respectively. Types F and G are similar structures wherein multiple slits are arranged in an arc shape, the slit width is 20 μm, and the pitches are 250 μm and 500 μm, respectively. In structures with these arc-shaped slits, the slit portions are deflected by the fluid force received and deform into shapes similar to that of an insect net. Metallic structures that deform into three-dimensional shapes of Types A–C and E–G have elastic structures and are designed to return to their original shapes when the hydrodynamic force is unloaded, even when deformed in response to hydrodynamic forces. Types D and H are structures with fine pores on flat plates, having diameters of 10 μm and 20 μm and pitches of 20 μm and 40 μm, respectively. These structures were fabricated to demonstrate that the three-dimensional deformation of a metal structure requires machining and design ingenuity.

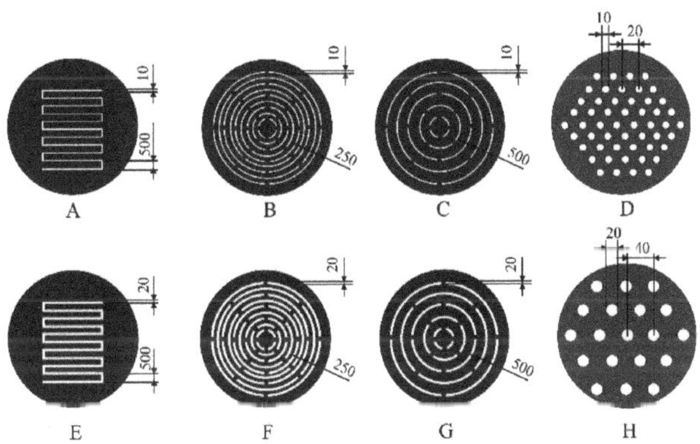

Figure 1. Overview of metal structure designs. The metal structures were designed with diameters of 15 mm and slit widths of 10 μm and 20 μm. Types A and E are comb-shaped, and Types B, C, F, and G are structures with multiple slits arranged in an arc shape. Types A–C and E–G have elastic structures and are designed to deform into three-dimensional shapes due to hydrodynamic force. The reason for utilizing different slit widths of 10 μm and 20 μm was to confirm that the thin film could be fabricated as the designed structure. The structures with fine pores on flat plates in Types D and H were fabricated to demonstrate the designs that needed to be devised.

2.2. Methods for Evaluating Physical Properties of Metal Structures

The stiffness of the fabricated material must be adjusted to fabricate a three-dimensionally deformable metal structure. It is important to consider the stiffness of the Ni structure for device design and performance, as it changes the shape maintenance and deformation capacities of the Ni structure. In addition, the film thickness, hardness, and slit width must be homogeneous. In Ni electroforming, the current is concentrated at the edge and outer peripheral portion of the substrate, owing to the current distribution problem, resulting in a thicker film [33]. Therefore, the film-thickness variation within the substrate plane was evaluated with four different distances, d = 24, 48, 60, and 84 mm, between the cathode and anode. The film thickness was measured at 89 locations within the substrate surface using a fluorescent X-ray film-thickness meter (SFT-9400, SII Nanotechnology Co., Ltd., Chiba, Japan). Furthermore, to evaluate the stiffness of the metal structures, the hardness of four types of metal structures was measured by adjusting the material with a film thickness of 40 μm. For the hardness measurements, a microhardness tester (HM-221, Mitutoyo Corporation, Kawasaki, Japan) was used to obtain measurements at nine points under a load of 0.49 N and a loading time of 10 s for determining the mean value. Furthermore, for the dimensions of the slit width of the fabricated metallic structure, a digital microscope (VHX-6000, KEYENCE Co., Ltd., Osaka, Japan) was used to measure the 12 in-plane locations for evaluating the variations.

2.3. Metal Structure Fabrication Method

Figure 2 shows the electroforming apparatus used to fabricate the metal structures. A Cu plate for a 4-inch silicon wafer (Yamamoto Plating Tester Co., Ltd., Tokyo, Japan) was employed as the cathode, and a Ni anode plate containing sulfur (Yamamoto Plating Tester Co., Ltd.) was utilized as the anode. A YPP-15,031 (Yamamoto Plating Tester Co., Ltd.) was used as the rectifier for the electroforming. Table 1 lists the basic configuration of the electroforming bath used in this study; the concentration of nickel sulfamate tetrahydrate in the bath (nickel sulfamate 900, JX Metal Shoji Co., Ltd., Tokyo, Japan) was 600 g/dm^3. The boric acid in the bath (Nacalai Tesque Inc., Kyoto, Japan) was prepared at a concentration of 30 g/dm^3. The pH of the electroforming bath was 4.0 ± 0.1, and the temperature of the electroforming bath was 40 ± 2 °C. The temperature was controlled by a temperature-control unit. The anode and cathode areas were 165 cm^2 and 136 cm^2, respectively. Ni dissolves on the anode side and Ni is deposited on the substrate on the cathode side. Moreover, Ni metal is deposited, and a Ni electroformed structure is formed. The chemical equations for the anode and cathode sides are shown in Equations (1)–(4) and (5)–(6), respectively.

$$Ni \rightarrow Ni^{2+} + 2e^- \tag{1}$$

$$Ni + H_2O \rightarrow NiO + 2H^+ + 2e^- \tag{2}$$

$$3Ni + 4H_2O \rightarrow Ni_3O_4 + 8H^+ + 8e^- \tag{3}$$

$$2H_2O \rightarrow O_2 + 4H^+ + 4e^- \tag{4}$$

$$Ni^{2+} + 2e^- \rightarrow Ni \tag{5}$$

$$2H^+ + 2e^- \rightarrow H_2 \tag{6}$$

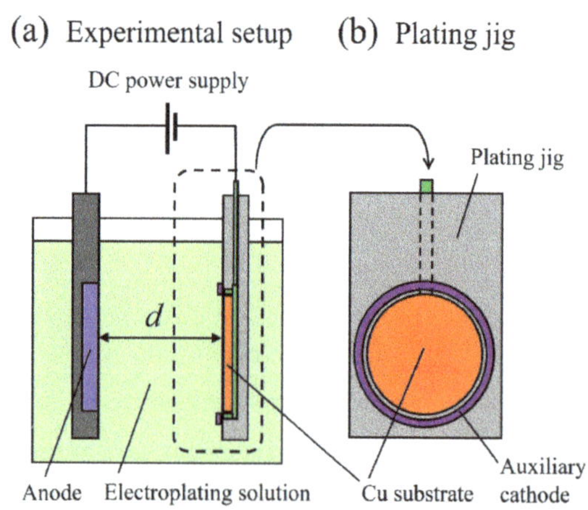

Figure 2. (**a**) is Experimental setup for electroforming. A Cu plate was used for the cathode, and a Ni anode plate containing sulfur was used for the anode, while d is the distance between the cathode and anode. (**b**) is Plating jig and the auxiliary cathode was installed at the periphery of the facility to ensure uniform film thickness.

Table 1. Bath composition for Ni electroforming.

$Ni(NH_2SO_3)_2{\cdot}4H_2O$	600 g/dm^3
H_3BO_3	30 g/dm^3
pH	4 ± 0.1
Temperature	40 ± 2 °C
Current density	5 A/dm^2
Paddle agitation	Stroke 100 mm, 60 rpm

3. Results and Discussion

3.1. Evaluation of Fabricated Substrate in-Plane Film Thickness

To evaluate the in-plane film thickness variation of the substrate produced by Ni electroforming, we measured the film thickness by setting a target film thickness of 10 µm and processing it for 10 min. Subsequently, the inter-electrode distance was set to the four aforementioned distances, d = 24, 48, 60, and 84 mm. Figure 3 shows the relationship between the inter-electrode distance and the electron-coating thickness. The longer the inter-electrode distance, the thinner is the film produced by Ni electroforming, and the average film thickness is the smallest at 6.7 µm with d = 84 mm. The variations are at their smallest with d = 48 mm, with an average film thickness of 12.9 µm and a standard deviation σ of 0.49 µm. Figure 4 shows the film-thickness distribution at each interelectrode distance. Negligible differences are observed in substrate in-plane film thickness, and the film-thickness variation is the smallest when d = 48 mm. With electroforming, the current is concentrated in the outer peripheral portion of the substrate when the distance between electrodes is large, and the current concentration in the central portion is high when the inter-electrode distance is small [34]. This suggests that d = 48 mm is suitable for achieving uniform film thickness. For d = 24, 48, and 60 mm, film thickness was more than 10 µm because only a small amount of electroforming was deposited on the auxiliary cathode installed at the periphery of the facility to ensure uniform film thickness, and the inside of the substrate became thick. Furthermore, because the electroforming is proportional to the electroforming time [35,36], if the throwing power is good, the desired film thickness may be expected. Based on these results, to suppress film-thickness variation in the various metal structures arranged on the substrate, we applied d = 48 mm in the subsequent evaluation. This is because the resulting coating thickness at the outermost peripheral portion being thin shows the interpolated numerical values of the unplated area and adjacent portion at the electroforming contact point, so can be ignored.

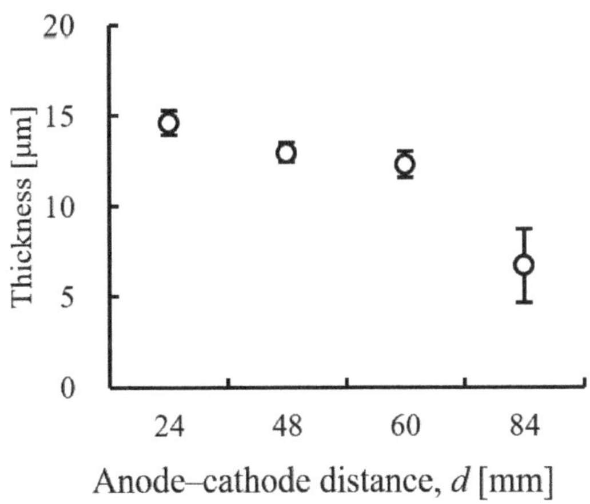

Figure 3. Evaluation of variations in Ni-coating thickness. The inter-electrode distance during Ni electroforming was set to four values, and the film thickness was measured by performing processing

for 10 min at a target film thickness of 10 μm. The average film thickness of 12.9 μm and σ = 0.49 μm minimized the variation at an interelectrode distance of 48 mm. As the coating thickness is proportional to the electroforming time, if the variation is small, the desired film thickness can be obtained via time adjustment.

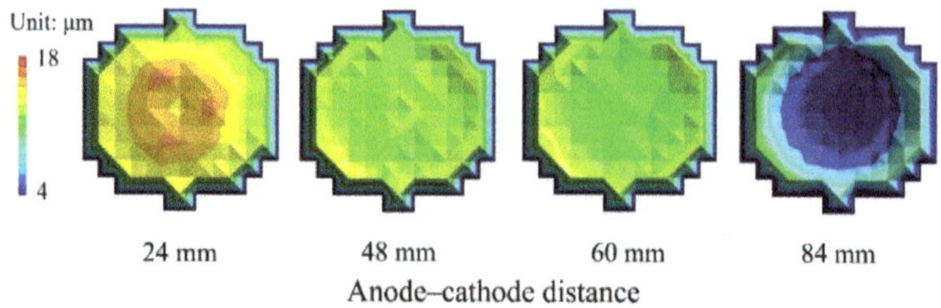

Figure 4. Substrate in-plane-coating thickness distribution. The average film thickness is 12.3 μm and σ = 0.72 μm, with a 60 mm inter-electrode distance, and at 48 mm, the variation is minimized, with an average film thickness of 12.9 μm and σ = 0.49 μm. There is almost no difference in the substrate in-plane film thickness with a 48 mm inter-electrode distance, and the variation is the smallest in this case.

3.2. *Evaluation of Stiffness of Metal Structures*

Figure 5 shows the hardness measurement results for the four types of samples, to evaluate the stiffness of the Ni structure. Sample (a) is a Ni coating obtained from the basic bath in Table 1, (b) is a coating adjusted with a high-hardness additive, (c) is a Ni–P alloy coating, and (d) is a sample of Ni–P coating heat-treated at 350 °C for 1 h. Compared to the Ni films in samples (a) and (b), the Ni–P alloy films in samples (c) and (d) are harder, especially in the sample (d), which was heat-treated and had a high Vickers hardness of 950 HV. This high hardness is thought to be due to crystallization [37,38] of the Ni–P alloy by heat treatment. However, we found that the samples (c) and (d) were much more difficult to deform than the other samples [39], and the coating cracked from simple bending, making them unsuitable as three-dimensional elastically deformable structures (Supplementary Materials Figure S1). Further, while sample (a) was easily deformed at approximately 300 HV, good stiffness with 600 HV hardness and elastic deformation were achieved when the plated coating was adjusted with a high curing additive in the sample (b); hence, we used this sample in subsequent evaluations. The Young's modulus of the sample (b) was 200 GPa. Figure 6 shows a schematic of the deformation expected when a metallic structure is subjected to hydrodynamic forces. When a solution is poured into a metal structure at a constant rate, hydrodynamic forces cause the metal structure to deform elastically into a certain shape. When the hydrodynamic force increases, the deformation of the metal structure increases and returns to normal when the hydrodynamic force is removed.

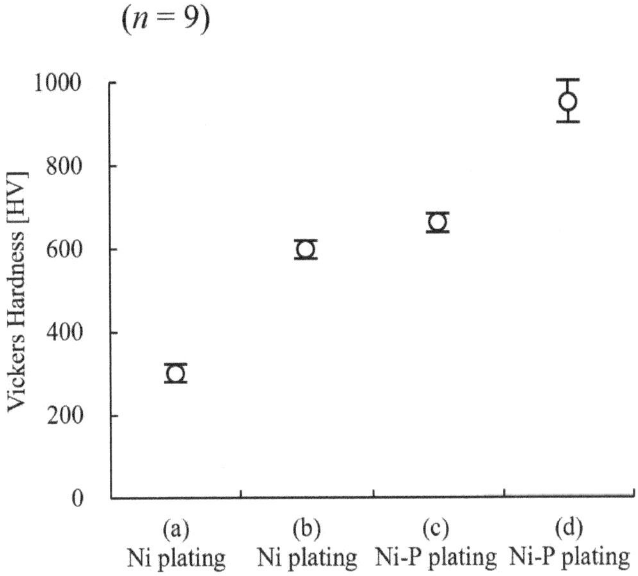

Figure 5. Ni-coating-hardness measurement results. Sample (**a**) is a Ni coating obtained from a basic bath, (**b**) is a coating adjusted by adding 5 mL/L [40] of sulfurous additive to the basic bath, (**c**) is a coating adjusted by adding 20 g/L of phosphorus acid to the basic bath, and (**d**) is a coating of the sample (c) heat-treated at 350 °C for 1 h. Sample (**b**) has a hardness of 600 HV and achieves stiffness allowing elastic deformation.

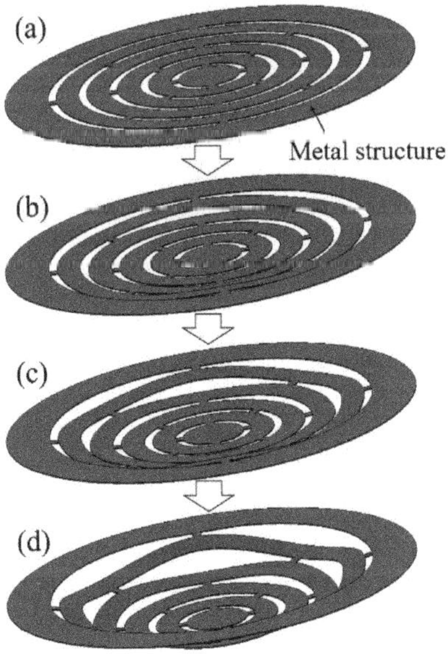

Figure 6. Schematic of deformation of metal structures due to hydrodynamic forces. (**a**) Initial state of the metal structure. (**b**) When the hydrodynamic forces are small, the deformation of the metal structure is small. (**c**,**d**) The greater the hydrodynamic forces is, the greater the deformation of the metal structure.

3.3. Fabrication of Three-Dimensionally Deformable Structures by Precision Electroforming

Ni structures are intended to be applied to filters. It is possible to perform various chemical modifications on the surface of the Ni structure by adding Au plating, which can add functions to the structural surface. Typical examples include the modification of self-assembled monolayers (SAMs), frequently used in analytical chemistry and biochemistry, and antibody modifications for antigen-antibody interactions. Figure 7 shows the fabrication process of a metal structure with a fine pattern. In the photolithography process, a 30 μm positive liquid resist (AZ P4903, Merck Performance Materials Co., Ltd., Tokyo, Japan) was applied to a Cu substrate by a spin coater (MS-A150, Mikasa Co., Ltd.) (Figure 7a) and was exposed to light at 1.75 J/cm^2 using a mask aligner (MA-20, Mikasa Co., Ltd.) (Figure 7b). A resist structure to serve as a mold was developed using an AZ 400 K Developer (Merck Performance Materials Co., Ltd.) (Figure 7c). To release the Ni structure from the Cu substrate, chromate treatment (2 g/dm^3, 2 min, 30 °C) with potassium dichromate solution was applied on the Cu substrate before Ni electroforming. Next, using an electroforming bath, we fixed a Ni plate containing sulfur to the anode and a fabricated resist structure to the cathode to form a Ni structure by electroforming (Figure 7d). The resist structure was released using stripping solution (AZ Remover 700, Merck Performance Materials Co., Ltd.). By removing the resist structure, we obtained a metal structure of the desired shape by releasing the Ni structure from the Cu substrate with tweezers (Figure 7e). Figure 8 shows the improvements in contact exposure in photolithography. As the liquid resist used has a high viscosity of 1550 mm^2/s, the resist was approximately 60 μm thick in the outer peripheral portion of the substrate post resist coating. The resist was physically removed using a 5 mm wide and 1 mm thick stainless-steel stick within 3 mm of the substrate edge, and the resist thickness was uniform at 30 μm. The removal of the excessively thick resist eliminated the gap between the mask and substrate, and the variation of the slit width decreased. Figure 9 shows photographs of the Ni structure fabricated by this process. We found eight types of structures, Types A–H, could be fabricated without defects, even in the slit portion or overall surface. Figure 10 shows the dimensional measurement results for the slit width of the Ni structure. By optimizing the photolithography conditions, the variation of the slit width improved from σ = 3.74 to σ = 0.25 μm. The slit width of Types A–D was 7.9 μm on average, with a design value of 10 μm, and the slit width of Types E–H was 18.3 μm on average, with a design value of 20 μm. In addition, the variation for all eight types was σ = 0.25–0.37 μm, and the fabrication was found to be precise. Although the slit width was approximately 2 μm smaller than the target value, it was expected that the desired slit width could be obtained by adjusting the design value. When fabricating a microstructure, the fact that the slit width becomes smaller than the target value is an advantage that may be adopted for various applications. The metal structure was fixed in the device shown in Figure 11, and hydrodynamic force was applied to demonstrate that the structure can deform from a two- to three-dimensional shape. Pressure was applied to the metal structure in a 9 mm diameter area around the center. Figure 12 presents a photograph of when a hydrodynamic force is applied to structures with a Type A comb shape and a Type B arc shape with multiple slits. The deformation amount of the Ni structures fabricated by this process changes according to the magnitude of the hydrodynamic force, and the structures return to their original shapes when the applied force is removed. Furthermore, if there is continuous flow at a constant rate, it is possible to maintain a constant deformation and the shape of the fluid. Incidentally, Types D and H, which have micropores formed in-plane, are self-supporting structures similar to the other structures. However, they do not deform into three-dimensional shapes under hydrodynamic forces.

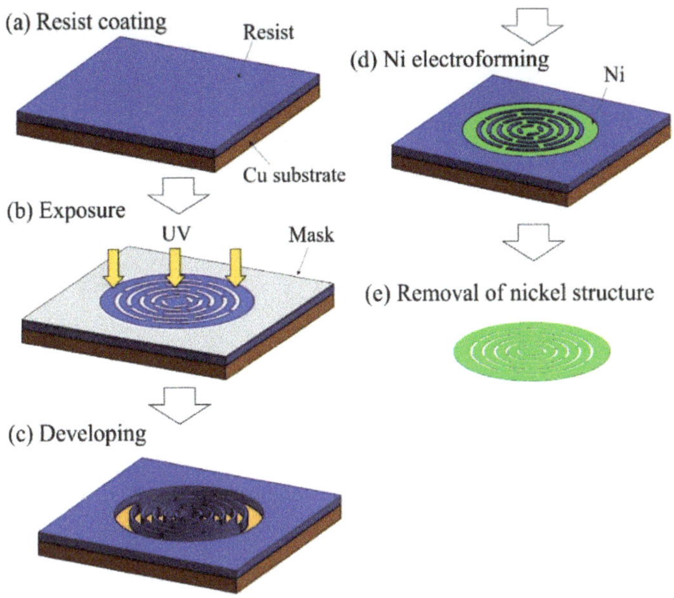

Figure 7. Schematic of the precision electroforming process. (**a**) A photoresist is applied on a Cu substrate. (**b**) Curing of photoresist by ultraviolet irradiation. (**c**) Mold creation by developing the photoresist. (**d**) Fabrication of Ni metal structures on Cu substrates by Ni electroforming. (**e**) Removal of the Ni structure to obtain a metal structure of the desired shape.

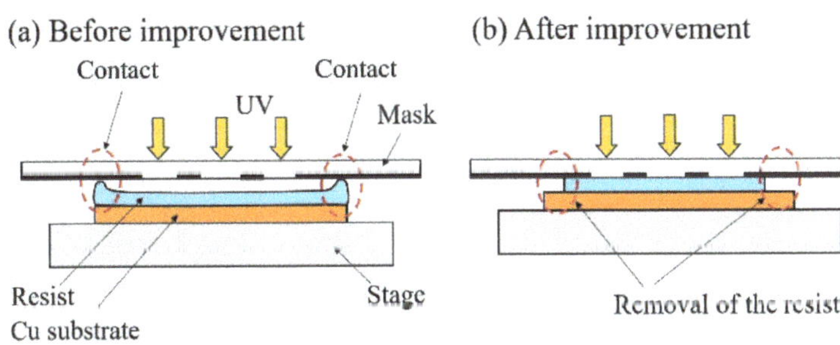

Figure 8. Improvement of the photolithography process. (**a**) After resist coating, the resist at the edge of the substrate is thick at approximately 60 μm. (**b**) Resist at the edges is physically removed after improvement. The thickness of the resist is uniform at 30 μm. There is no gap between the mask and substrate in the contact exposure.

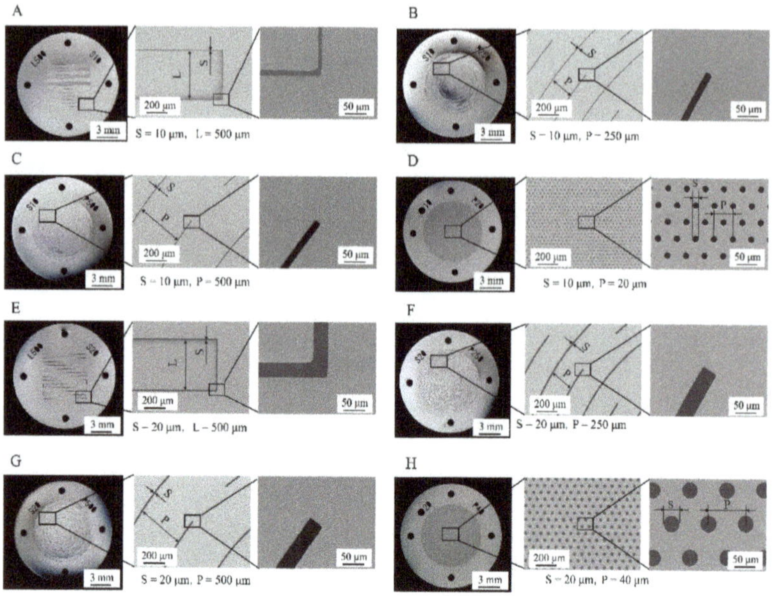

Figure 9. Photographs of the fabricated Ni structures. Eight types of Ni structures were fabricated precisely. Metal structures that deform three-dimensionally of Types (**A–C**) and (**E–G**) return to the original state even when deformed by hydrodynamic force. Metal structures of Types (**D**) and (**H**) are formed using the same processes and materials as metal structures that deform three-dimensionally and do not deform into three-dimensional shapes. All types appear warped under the light from the digital microscope, but the Ni structures are not warped after electroforming.

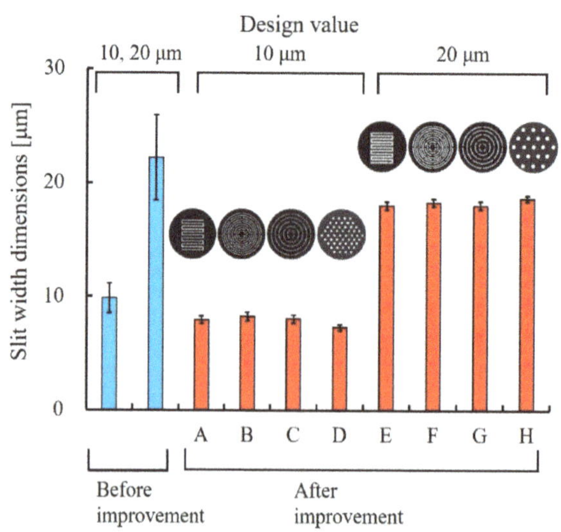

Figure 10. Metal-structure slit-width dimensions before and after improvement of the photolithography process. The dimensional variation decreases from $\sigma = 3.74$ to $\sigma = 0.25$ μm as a result of optimizing the photolithography conditions. The desired slit width is expected if the design value is adjusted.

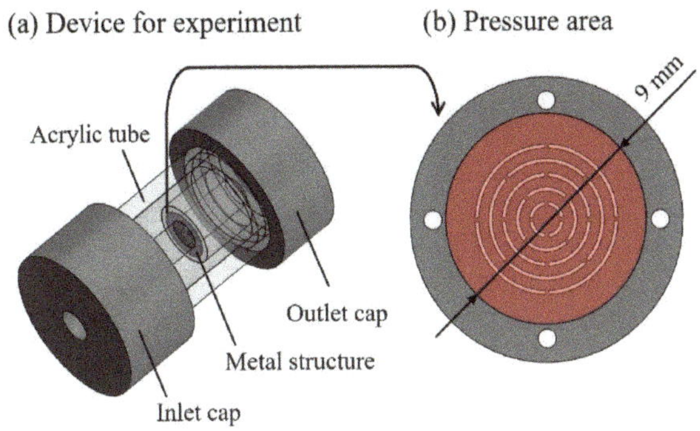

Figure 11. Overview of the device for the experiment. (**a**) The metal structure was fixed inside the device, and the hydrodynamic force was applied. The metal structure was fixed to an acrylic tube with an inner diameter of 9 mm. (**b**) Pressure was applied to the metal structure in a 9-mm-diameter area around the center.

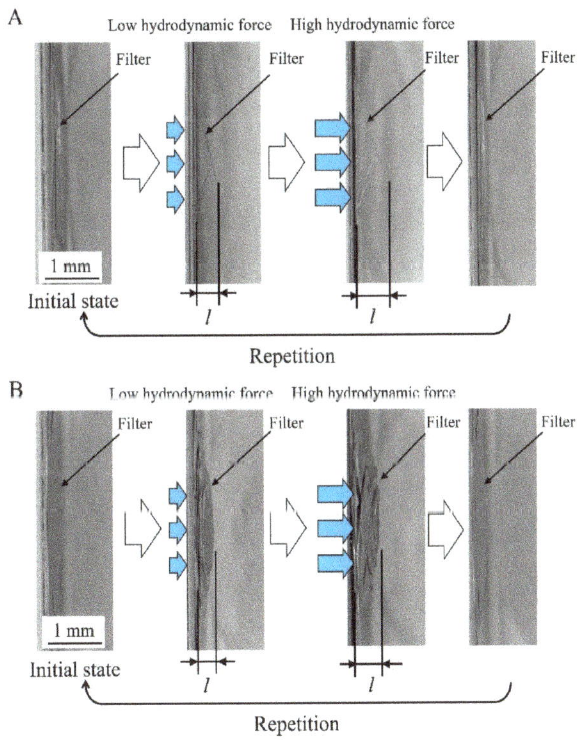

Figure 12. Photographs of three-dimensionally deformable metal structures. Type A is comb-shaped, and Type B is a metal structure with multiple arc-shaped slits. The deformation quantity l changes with the magnitude of the hydrodynamic force. When the hydrodynamic force is unloaded, the structure returns to its original state. The deformation of the Type A structure at a flow rate of 300 mL/h was approximately 110 μm, and that at a flow rate of 2400 mL/h was approximately 820 μm. The deformation of the Type B structure at a flow rate of 300 mL/h was approximately 40 μm, and that at a flow rate of 2400 mL/h was approximately 90 μm. The deformations of both

structures increased linearly with increasing flow rate. In addition, a comparison of the simulated and measured deformation of the two metal structures under a flow rate of 300 mL/h revealed that Type A showed deformation of approximately 120 μm (simulation value), and the measurement result was 110 μm, while Type B showed deformation of approximately 40 μm (simulation value), and the measurement result was 40 μm. Under low flow-rate conditions, the simulated and measured results showed good agreement (Supplementary Materials Figure S2). There was no plastic deformation, even under conditions with a flow rate of 2400 mL/h. As there were no realistic conditions for exceeding 2400 mL/h, it was not possible to verify until plastic deformation occurred. In this study, as a limitation to elastic deformation, results were shown for conditions with a flow rate of 300–2400 mL/h.

Figure 13 shows a cross-sectional scanning-electron microscopy (SEM) image of a Type B metal structure fabricated using 10 μm and 2 μm as the target film thicknesses for Ni electroforming and Au plating, respectively. Type B is obtained with Au coating from the basic baths in Table 2. The digital microscopy and cross-sectional SEM images show that there are no pinholes or cracks on the surface of the metal structure, demonstrating the absence of structural defects. The Ni electroforming thickness is 10.1 μm, the Au-plating thickness is 2 μm, and the film thickness can be formed as designed. In addition, a structure with a high aspect ratio (film thickness/slit width = 3.9) could be fabricated with good precision using this process. Incidentally, the yield of the fabricated structure is 90%. The three-dimensionally deformable microstructure fabricated using the process developed in this study can be applied to bioMEMSs, analysis chips, and μTASs from the perspective of size. Therefore, these structures may be expected to be applied in the biomedical field.

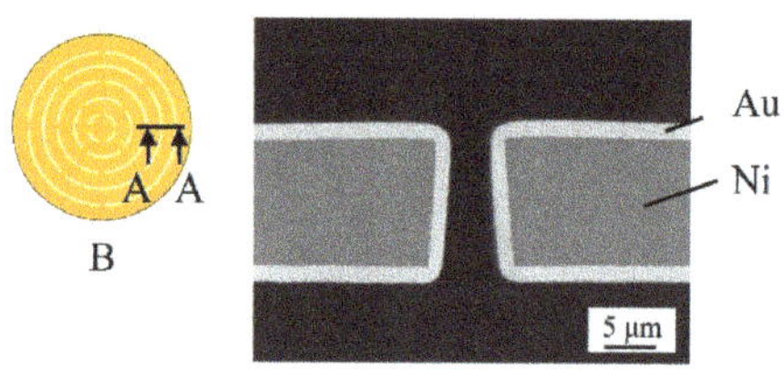

Figure 13. Cross-sectional photograph of a Type B three-dimensionally deformable metal structure (slit width: 10 μm) fabricated with a Ni electroforming thickness of 10 μm and Au-plating thickness of 2 μm. The tapered shape is due to the effect of diffraction of light during exposure.

Table 2. Bath composition for Au plating.

$Na_3Au(SO_3)_2$	21.6 g/dm^3
Na_2SO_3	75.6 g/dm^3
pH	8 ± 0.1
Temperature	45 ± 2 °C
Current density	0.2 A/dm^2
Agitation	Stirring 300 rpm

4. Conclusions

The objectives of this study were to fabricate a metal structure that can be elastically deformed into a three-dimensional shape in response to hydrodynamic forces and to propose a fine pattern fabrication approach using precision electroforming. Physical properties, such as the film thickness, hardness, and slit width, were evaluated to fabricate the fine metal structures using this technique. Based on the results, we achieved $\sigma = 0.49$ μm with an average film thickness of 12.9 μm and $\sigma = 0.25$ μm with a hardness of 600 HV and slit width of 7.9 μm. The conditions for fabricating precision metal structures were also evaluated. When the intended metal structure was fabricated using the derived conditions, we successfully fabricated metal structures capable of dynamically deforming into three-dimensional shapes according to the hydrodynamic force. In this study, structures with hardness ranging 300–950 HV and flow velocities of 300–2400 mL/h were investigated, but they were within the elastic deformation range and did not deform plastically. As the type of metal and physical properties of metal structures fabricated using this technique can be varied, these structures can be applied to various fields and applications such as bioMEMSs and μTASs in the future.

Supplementary Materials: The following supporting information can be downloaded at: https://www.mdpi.com/article/10.3390/mi13071046/s1. Figure S1: Simplified flexural evaluation of four types of Ni film samples (a)–(d). The Ni films in (a) and (b) were found to deform elastically and return to their original shapes without damage. The samples with Ni-P alloys in (c) and (d) cracked and broke while bending. In addition, they did not return to their original shapes. Figure S2: Simulation analysis results for the deformation of the Type A and B metal structures. The conditions used in this analysis are summarized in Table S1. The measured deformation and simulation results were compared. Under a flow rate of 300 mL/h, Type A showed a deformation (simulation value) of approximately 120 μm, and the measurement result was 110 μm, whereas Type B showed a deformation (simulation value) of approximately 40 μm, and the measurement result was 40 μm. Under a flow rate of 2400 mL/h, Type A showed a deformation (simulation value) of approximately 1080 μm, and the measurement result was 820 μm, whereas Type B showed a deformation (simulation value) of approximately 280 μm, and the measurement result was 90 μm. Thus, the simulation and measurement results agree well under the condition of a low flow rate. Table S1: Deformation analysis conditions.

Author Contributions: Conceptualization, Y.N. (Yuta Nakashima); methodology, Y.K., M.I., H.B., T.I., Y.N. (Yoshitaka Nakanishi) and Y.N. (Yuta Nakashima); validation, S.K. and K.Y.; formal analysis, S.K.; investigation, S.K., S.F., S.N. and Y.N. (Yuta Nakashima); resources, S.K., K.Y., Y.K., M.I., H.B., T.I., Y.N. (Yoshitaka Nakanishi) and Y.N. (Yuta Nakashima); writing—original draft preparation, S.K. and Y.N. (Yuta Nakashima); writing—review and editing, Y.N. (Yuta Nakashima); project administration, Y.N. (Yuta Nakashima); supervision, Y.N. (Yuta Nakashima); project administration, Y.N. (Yuta Nakashima); funding acquisition, Y.N. (Yuta Nakashima). All authors have read and agreed to the published version of the manuscript.

Funding: This work was partly supported by a fund from the Japan Society for the Promotion of Science (JSPS) KAKENHI, via the Grant-in-Aid for Young Scientists (A) 16H06078 and the Ministry of Economy, Trade and Industry (METI) under projects to support the advancement of strategic core technologies, 17939048. The funding sources were not involved in the study design, concept, data collection, analysis, interpretation, manuscript preparation, or submission.

Institutional Review Board Statement: Not applicable.

Informed Consent Statement: Not applicable.

Data Availability Statement: All the data generated or analyzed during this study are included in this published article.

Conflicts of Interest: The authors declare that they have no conflict of interest.

References

1. Kamilla, S.K.; Ojha, M. Review on nano-electro-mechanical-system devices. *Mater. Today Proc.* **2021**. [CrossRef]
2. Eshkalak, S.K.; Ghomi, E.R.; Dai, Y.; Choudhury, D.; Ramakrishna, S. The role of three-dimensional printing in healthcare and medicine. *Mater. Des.* **2020**, *194*, 108940. [CrossRef]
3. Esfahani, R.N.; Shuttleworth, M.P.; Doychinov, V.; Wilkinson, N.J.; Hinton, J.; Jones, T.D.A.; Ryspayeva, A.; Robertson, I.D.; Marques-Hueso, J.; Desmulliez, M.P.Y.; et al. Light based synthesis of metallic nanoparticles on surface-modified 3D printed substrates for high performance electronic systems. *Addit. Manuf.* **2020**, *34*, 101367. [CrossRef]
4. Arzt, E.; Quan, H.; McMeeking, R.M.; Hensel, R. Functional surface microstructures inspired by nature—From adhesion and wetting principles to sustainable new devices. *Prog. Mater. Sci.* **2021**, *119*, 100778. [CrossRef]
5. Shen, L.; Li, Y.; Zhong, W.; Wu, J.; Cheng, J.; Jin, L.; Hu, X.; Ling, Z. Fabrication of micro/nanoporous templates with a novel hierarchical structure by anodization of a patterned aluminum surface. *Electrochem. Commun.* **2021**, *126*, 107014. [CrossRef]

6. Shams, H.; Basit, K.; Khan, M.A.; Saleem, S.; Mansoor, A. Realizing surface amphiphobicity using 3D printing techniques: A critical move towards manufacturing low-cost reentrant geometries. *Addit. Manuf.* **2021**, *38*, 101777. [CrossRef]
7. Greenwood, P.A.; Greenway, G.M. Sample manipulation in micro total analytical systems. *Trends Analyt. Chem.* **2002**, *21*, 726–740. [CrossRef]
8. Bagherabadi, K.M.; Sani, M.; Saidi, M.S. Enhancing active electro-kinetic micro-mixer efficiency by introducing vertical electrodes and modifying chamber aspect ratio. *Chem. Eng. Process.* **2019**, *142*, 107560. [CrossRef]
9. Satoh, W.; Shimizu, Y.; Kaneto, T.; Suzuki, H. On-chip microfluidic transport and bio/chemical sensing based on electrochemical bubble formation. *Sens. Actuators B Chem.* **2007**, *123*, 1153–1160. [CrossRef]
10. Erickson, D.; Li, D. Integrated microfluidic devices. *Anal. Chim. Acta* **2004**, *507*, 11–26. [CrossRef]
11. Prakash, B.; Singh, A.; Katoch, V.; Sharma, M.; Panda, J.J.; Sharma, J.; Ganguli, A.K. Flow synthesis and in-channel photocatalysis of antimicrobially active ZnS quantum dots using an efficient planar PMMA microreactor. *Nano Express.* **2020**, *1*, 030030. [CrossRef]
12. Nakashima, Y.; Hata, S.; Yasuda, T. Blood plasma separation and extraction from a minute amount of blood using dielectrophoretic and capillary forces. *Sens. Actuators B Chem.* **2010**, *145*, 561–569. [CrossRef]
13. Nagai, M.; Kato, K.; Shibata, T. Underwater motion of hydrogel microstructure by optofluidic lithography studied with gap control and object holding platform. *Microelectron. Eng.* **2016**, *164*, 108–114. [CrossRef]
14. Arrazola, P.J.; Rech, J.; M'Saoubi, R.; Axinte, D. Broaching: Cutting tools and machine tools for manufacturing high quality features in components. *CIRP Ann.* **2020**, *69*, 554–577. [CrossRef]
15. Yang, B.; Zhang, G.; Ran, Y.; Yu, H. Kinematic modeling and machining precision analysis of multi-axis CNC machine tools based on screw theory. *Mech. Mach. Theory* **2019**, *140*, 538–552. [CrossRef]
16. Khamar, P.; Prakash, S. Investigation of dimensional accuracy in CO_2 laser cutting of PMMA. *Mater. Today Proc.* **2020**, *28*, 2381–2386. [CrossRef]
17. Matsuura, T.; Chabloz, M.; Jiao, J.; Yoshida, Y.; Tsutsumi, K. A method to evade silicon backside damage in deep reactive ion etching for anodically bonded glass–silicon structures. *Sens. Actuators A Phys.* **2001**, *89*, 71–75. [CrossRef]
18. Miki, N.; Teo, C.J.; Ho, L.C.; Zhang, X. Enhancement of rotordynamic performance of high-speed micro-rotors for power MEMS applications by precision deep reactive ion etching. *Sens. Actuators A Phys.* **2003**, *104*, 263–267. [CrossRef]
19. Waits, C.M.; Morgan, B.; Kastantin, M.; Ghodssi, R. Microfabrication of 3D silicon MEMS structures using gray-scale lithography and deep reactive ion etching. *Sens. Actuators A Phys.* **2005**, *119*, 245–253. [CrossRef]
20. Le, N.N.; Huynh, K.K.; Phan, T.C.H.; Dang, T.M.D.; Dang, M.C. Fabrication of 25μm-filter microfluidic chip on silicon substrate. *Adv. Nat. Sci. Nanosci. Nanotechnol.* **2017**, *8*, 015003. [CrossRef]
21. Lou, J.; Shrotriya, P.; Allameh, S.; Buchheit, T.; Soboyejo, W.O. Strain gradient plasticity length scale parameters for LIGA Ni MEMs thin films. *Mater. Sci. Eng. A* **2006**, *441*, 299–307. [CrossRef]
22. Zhu, X.; Liu, G.; Xiong, Y.; Guo, Y.; Tian, Y. Fabrication of PMMA microchip of capillary electrophoresis by optimized UV-LIGA process. *J. Phys. Conf. Ser.* **2006**, *34*, 875–879. [CrossRef]
23. Malek, C.K.; Saile, V. Applications of LIGA technology to precision manufacturing of high-aspect-ratio micro-components and -systems: A review. *Microelectron. J.* **2004**, *35*, 131–143. [CrossRef]
24. Ma, Y.; Liu, W.; Liu, C. Research on the process of fabricating a multi-layer metal micro-structure based on UV-LIGA overlay technology. *Nanotechnol. Prec. Eng.* **2019**, *2*, 83–88. [CrossRef]
25. McGeough, J.A.; Leu, M.C.; Rajurkar, K.P.; De Silva, A.K.M.; Liu, Q. Electroforming process and application to micro/macro manufacturing. *CIRP Ann.* **2001**, *50*, 499–514. [CrossRef]
26. Chen, X.; Wei, X.; Jiang, K. Fabrication of large-area nickel nanobump arrays. *Microelectron. Eng.* **2009**, *86*, 871–873. [CrossRef]
27. Chen, X.; Liu, L.; He, J.; Zuo, F.; Guo, Z. Fabrication of a metal micro mold by using pulse micro electroforming. *Micromachines* **2008**, *9*, 203. [CrossRef]
28. Clausi, D.; Gradin, H.; Braun, S.; Peirs, J.; Stemme, G.; Reynaerts, D.; van der Wijngaart, W. Robust actuation of silicon MEMS using SMA wires integrated at wafer-level by nickel electroplating. *Sens. Actuator A Phys.* **2013**, *189*, 108–116. [CrossRef]
29. Hou, J.; Zhang, H.; Zhang, N.; Fang, F. Characterization of manufacturability of microstructures for micro-injection moulding of micro devices using star patterns. *J. Micromech. Microeng.* **2020**, *30*, 025001. [CrossRef]
30. Akgönüllü, S.; Bakhshpour, M.; Pişkin, A.K.; Denizli, A. Microfluidic Systems for Cancer Diagnosis and Applications. *Micromachines* **2021**, *12*, 1349. [CrossRef]
31. Zagatto, E.A.; Rocha, F.R. The multiple facets of flow analysis. A tutorial. *Anal. Chim. Acta* **2019**, *1093*, 75–85. [CrossRef] [PubMed]
32. Guijt, R.; Manz, A. Miniaturised total chemical-analysis systems (μTAS) that periodically convert chemical into electronic information. *Sens. Actuators B Chem.* **2018**, *273*, 1334–1345. [CrossRef]
33. Yang, H.; Kang, S.-W. Improvement of thickness uniformity in nickel electroforming for the LIGA process. *Int. J. Mach. Tools Manuf.* **2000**, *40*, 1065–1072. [CrossRef]
34. Sun, Y.T.A.; Liu, M.-C.; Zhang, Y.; Huang, Y.-P.; Chen, W.-E. Development of an electrolyte jet type apparatus for manufacturing electroplated diamond wires. *Precis. Eng.* **2021**, *68*, 351–357. [CrossRef]
35. Yamamoto, T.; Igawa, K.; Tang, H.; Chen, C.-Y.; Chang, T.-F.M.; Nagoshi, T.; Kudo, O.; Maeda, R.; Sone, M. Effects of current density on mechanical properties of electroplated nickel with high speed sulfamate bath. *Microelectron. Eng.* **2019**, *213*, 18–23. [CrossRef]
36. Kim, B.-H.; Park, B.-J.; Kim, J.B. Process effects of double step DRIE and Ni–Co electroforming for a trench-type cantilever probe for a fine-pitched MEMS probe card. *Sens. Actuators A Phys.* **2009**, *152*, 252–260. [CrossRef]

37. Zhan, X.; Ernst, F. Crystallization micro-mechanism of near-eutectic amorphous Ni–P. *Acta Mater.* **2016**, *104*, 274–282. [CrossRef]
38. Naderi, J.; Sarhan, A.A.D. Measure and evaluate the hardness of the electrodeposited nickel-phosphorous (Ni-P) thin film coating on carbon steel alloy for automotive applications. *Measurement* **2019**, *139*, 490–497. [CrossRef]
39. Lelevic, A.; Walsh, F.C. Electrodeposition of NiP alloy coatings: A review. *Surf. Coat. Tech.* **2019**, *369*, 198–220. [CrossRef]
40. Kanamori, G.; Yasuda, Y.; Watanabe, M.; Honma, H.; Takai, O. Evaluation of physical properties of sulfamate Ni-Mn electroplating film. *J. Surf. Finish. Soc. Jpn.* **2017**, *68*, 564–569. [CrossRef]

micromachines

MDPI

Article

Evaluation of Bronze Electrode in Electrical Discharge Coating Process for Copper Coating

JagadeeswaraRao Maddu [1], Buschaiah Karrolla [1], Riyaaz Uddien Shaik [2,3,*], Hassan Elahi [4] and Krishnaiah Arkanti [1]

1 Department of Mechanical Engineering, University College of Engineering, Osmania University, Hyderabad 500007, India
2 Department of Astronautics Electrical and Energy Engineering, University of Rome "La Sapienza", Via Eudossiana 18, 00184 Rome, Italy
3 Department of Civil and Environmental Engineering, University of California, Los Angeles, CA 90095, USA
4 Department of Mechanical and Aerospace Engineering, University of Rome "La Sapienza", Via Eudossiana 18, 00184 Rome, Italy
* Correspondence: riyaaz.shaik@uniroma1.it

Abstract: One of the widely used non-traditional machines for machining of hard materials into complex shapes and different sizes is the electrical discharge machine (EDM). Recently, the EDM has been used for deposition by controlling the input parameters (current and duty cycle). This work was carried out to evaluate the readily available bronze (88% Cu + 12% Sn) electrode for deposition of copper material on titanium alloy. Experiments were conducted according to Taguchi experimental design considering the input parameters of current, Ton, Toff and preheating temperature of substrates. Titanium alloy was further hardened by preheating at temperatures of 100 °C, 300 °C and 500 °C and quenching in brine, castor oil and vegetable oil in order to avoid workpiece erosion. After this treatment, hardness, grain area, grain diameter and number of grains were characterized to compare with pretreated substrates. Then, the treated substrates were taken for copper deposition with the EDM. Output parameters such as material deposition rate (MDR), electrode wear rate (EWR), coating thickness (CT), elemental composition and surface crack density (SCD) were found. Material characterization was carried out using a scanning electron microscope (SEM) with energy dispersive X-ray spectroscopy (EDX) and optical microscopy. Output parameters were optimized with technique for order of preference by similarity to ideal solution (TOPSIS) to find optimum parameters. A sixth experiment with parameter values of Ton of 440 μs, Toff of 200 μs, preheating temperature of 300 °C and quenching medium of castor oil was optimum with MDR of 0.00506 g/m, EWR of 0.00462 g/m, CT of 40.2 μm and SCD 19.4 × 10^7 μm^2.

Keywords: electrical discharge coating; microhardness; material deposition rate; TOPSIS

Citation: Maddu, J.; Karrolla, B.; Shaik, R.U.; Elahi, H.; Arkanti, K. Evaluation of Bronze Electrode in Electrical Discharge Coating Process for Copper Coating. *Micromachines* 2023, 14, 136. https://doi.org/10.3390/mi14010136

Academic Editor: Nikolaos Tapoglou

Received: 26 November 2022
Revised: 29 December 2022
Accepted: 30 December 2022
Published: 4 January 2023

1. Introduction

The EDM is a non-traditional machine that supports the fabrication of complex and intrinsic shapes with an excellent surface finish in materials [1]. This is a well-established technique in the fields of biomedical, automotive, chemical, aerospace, tool and die industries [2]. Usually, the EDM removes material by repeated sparks between the workpiece and tool electrode immersed in dielectric medium [3]. The thermal energy between electrode and workpiece creates high temperature plasma, which erodes, melts and evaporates the workpiece material [4]. Meanwhile, the EDC process requires a low current and high duty cycle which reverses the process of the EDM [5]. In EDC, electrode material is deposited on the workpiece with a difference in parameters [6]. Even in EDC, high frequency electrical discharges or sparks cause the workpiece material to melt and vaporize. Extreme temperatures in the range of 8000–12,000 °C lead to erosion and vaporization of workpiece and electrode [7]. Then, material transfer occurs from the electrode to the workpiece under the suitable process conditions and parameter setup. On the surface of the workpiece, a recast layer of redeposited melt materials from the electrode is deposited on the workpiece which is immersed in dielectric medium [8,9]. The deposited material solidifies and forms a coating in dielectric medium. This process modifies the workpiece surface by generating new compositions which can be further processed by quenching and hardening processes [10].

In this work, superalloy Ti6Al4V is exploited as a workpiece due to its essential characteristics, viz., fracture toughness, biocompatibility, improved ductility, wear resistance, yield strength and corrosion resistance [11]. This alloy has proven its applicability in various fields such as medical implants, marine appliances, airframes, automotive industry, etc. Among them, usage of this alloy in some applications, such as medical implants, wastewater treatment plants, etc., requires antibacterial coating [12]. Copper material has proven antibacterial activity since it helps to increase human immunity [13]. So, in this work, copper material was proposed as coating material.

In our previous research, attempts were made to coat copper on titanium alloy using copper electrodes. Firstly, an attempt was made using copper electrodes and it was sparsely coated on the workpiece [14]. Instead, workpiece material was removed and microhole formation was observed. Secondly, brass, which is an alloy of copper (67%) and zinc (33%), was selected to coat copper and a regular, crack free and stable coating of thickness of 22 μm was obtained. In this work, one more attempt is made with a bronze electrode which is also an alloy of copper containing from 0.5 to 11% tin and 0.01 to 0.35% phosphorus.

Bronzes or tin bronzes are alloys containing copper, tin and phosphorus. The addition of tin increases the corrosion resistance and strength of the alloy whereas phosphorus increases the wear resistance and stiffness of the alloy [6]. Phosphor bronzes have high fatigue resistance, solderability, excellent formability and high corrosion resistance. Phosphorus bronze has established applicability in sleeve bearings, cam followers, thrust washers and electrical products such as diaphragms, corrosion resistant bellows and spring washers [14]. This material has proven strength, high wear resistance, fatigue resistance with good machinability and corrosion resistance [15].

Researchers around the world are working to stabilize and standardize the procedure of electrical discharge coatings. Some of the examples are as follows: Algodi et al. [16] have examined the hardness variation of TiC-Fe nanostructured coating by varying the input parameters such as current and Ton and concluded in their study that the latter is the most influencing factor. Mussada et al. [17] have investigated the possibility of PM electrodes for EDM-based surface modification. The investigation was performed in a stepwise manner, though it takes more time, and a good surface finish was obtained. Hsu et al. [18] varied input parameters of the EDM, viz., material removal rate (MRR), surface roughness (Ra) and electrode wear rate (EWR), to improve the surface finishing. Here, oxygen plasma etching treatment was performed to decrease the surface roughness [19]. In order to further increase the surface characteristics, physical vapor deposition (PVD) was performed to coat TiN. Algodi et al. [20] investigated antibacterial coating on titanium alloy by mixing silver nanopowder with dielectric medium and compared it without mixing powder. It was concluded that electrode material deposition is comparatively less when dielectric medium is mixed with silver nanopowder. Tyagi et al. [21] conducted a study to coat a mild steel (MS) workpiece surface with WS2 and copper green compact electrodes in different composition mixing ratios. It was observed that WS2 increases coating thickness whereas current and duty factors influence wear and hardness. Murray et al. [22] reported their work varying the input parameters of EDC to coat different materials of copper, zirconium and tungsten carbide on stainless steel. Bui et al. [23] studied the elemental composition of the modified workpiece surface, tool electrode and dielectric fluid with immersed powder particles. Due to the application of titanium material (Ti6Al4V) in various fields, many studies are ongoing around the world. For instance, Wuyi Ming et al. [24] studied microporosity and microtrench machining, Kahlin et al. [25] studied fatigue behavior of materials, Zhen Zheng et al. [26] worked on laser-induced plasma micromachining and Schnell et al. [27] studied surface topography using femtosecond laser-induced periodic surface structures (FLIPSSs) and micrometric ripples (MRs).

In this work, a bronze electrode was selected to coat copper on titanium alloy (Ti6Al4V) to compare with our previous attempts. Prior to coating, workpiece substrates were preheated at different temperatures of 100 °C, 300 °C and 500 °C and quenched in brine, castor oil and vegetable oil in order to avoid workpiece erosion [28]. After this treatment, hardness, grain area, grain diameter and number of grains were characterized to compare with pretreated substrates. EDC input parameters selected to be optimized were current, Ton, Toff and preheating temperature. TOPSIS techniques were used to optimize the input parameters and material characterization was conducted using SEM with EDX. Explanations about the electrode and workpiece material are provided in Section 2. The experimental procedure is described in Section 3 with a process flowchart. Section 4 explains the results obtained from TOPSIS and material characterization. Section 5 concludes the paper with short conclusions on this work.

2. Materials and Methods

2.1. Workpiece and Electrode Materials

In this work, titanium alloy was selected as a workpiece due to its applications in various fields, mainly as medical implants, and a bronze electrode was selected as electrode material in order to evaluate it for copper coating [29]. Initially, a plate of titanium was obtained from Ramesh Steels Corporation Pvt. Ltd., Mumbai, India and then substrates of 20 mm × 20 mm × 8 mm were made using a wire-cut EDM, whereas bronze electrodes of 100 mm in length and 10 mm in diameter were made by power-hacksaw. EDM 30 was used as dielectric fluid in this experiment.

The chemical composition, density (kg/m^2), melting point (°C), specific heat capacity (J/g °C) and hardness of the electrode and substrate are shown in Table 1 [11].

Table 1. Physical and Chemical Composition of Electrode and Substrate.

Properties	Electrode	Substrate
Chemical composition	Cu 88% Sn 10% Zn 2%	C 0.08, Fe 0.25, Al 6, V 4, Ti balance
Density (kg/m^3)	8770	4.42
Melting point (°C)	1035	1878
Specific heat capacity (J/g °C)	370	553
Hardness	170	300

The three levels of EDM machining process parameters selected are shown in Table 2. Output parameters such as surface quality, surface topography and homogeneity of the coatings rely on the input process parameters, viz., current, Ton, Toff and temperature, as shown in Table 2 [19]. Taguchi L9 design was followed to prepare the combination of parameters [30,31] as shown in Table 2.

Table 2. EDM Input process parameters.

S. No.	Input Process Parameters	Level
1	Current (Amps)	4 8 12
2	Ton (μs)	280 360 440
3	Toff (μs)	200 300 400
4	Temperature (°C)	100 300 500

2.2. Output Process Parameters

In this work, output process parameters considered for optimization are material deposition rate (MDR) [32], electrode wear rate (EWR) [33,34] and surface crack density (SCD) [25]. MDR can be represented as

$$\text{MDR} = \frac{\text{WBM} - \text{WAM}}{\text{Time}} (\text{gram}/\text{min}) \tag{1}$$

where WAM = weight after machining and WBM = weight before machining.

EWR can be represented as

$$\text{EWR} = \frac{\text{EBM} - \text{EAM}}{\text{Time}} (\text{gram}/\text{min}) \tag{2}$$

where EAM = weight of electrode after machining and EBM = weight of electrode before machining.

Finally, surface crack density was considered which can represented as follows:

$$\text{SCD} = \frac{\text{T}_\text{l}}{\text{A}_\text{i}} \left(\mu\text{m}/\mu\text{m}^2\right) \tag{3}$$

where Tl is total crack length in μm and Ai is image area in μm^2.

Every researcher is interested in this parameter to provide crack free coating since it is the proper measure of cracks. This parameter depends upon the coefficient of thermal expansion of coating and workpiece material.

2.3. Methods

TOPSIS

The procedure for TOPSIS optimization is as follows:

Step 1: The first step is to create a decision matrix. This method consists of alternatives in the rows and attributes in the columns. The matrix format can be expressed as follows [19–35].

$$D = \; a1 \,.\, m \begin{bmatrix} X11 & \cdots & X1n \vdots \ddots \vdots Xm1 & \cdots & Xmn \end{bmatrix} \tag{4}$$

Here, a (i = 1,2,3, ... ,m) = all possible alternatives, x (j = 1,2,3, ... ,n) = the attributes related to performance of alternatives, j = 1,2,3, ... ,n and x_{ij} represents the performance of i with respect to attribute j.

Step 2: In this step, normalization of the above decision matrix is carried out and we obtain a normalized decision matrix γ_{ij}. The formula for r_{ij} is given below:

$$\gamma_{ij} = \frac{-x_{ij}}{\sqrt{\sum_{i=1}^{m} x_{ij}^2}} \tag{5}$$

Step 3: Here, weights are assigned according to the importance and the weighted normalized decision matrix can be calculated by using the formula V $= w_j \gamma_{ij}\, w \cdot r$.

$$V = \left[v_{ij}\right] \tag{6}$$

$$\sum_{j=1}^{n} w_j = 1 \tag{7}$$

Step 4: Positive and negative ideal solutions are calculatedtion is calculated by using the follow in this step. The solutions can be represented as the positive ideal (best) solution [36].

$$a^{+} = \left\{ (v_{ij}, j \in J) \left(v_{ij} J \in J \right) J \right\} \tag{8}$$

$$v = \left\{ v1^{+}, v2^{+}, v3^{+} \ldots\ldots .vj^{+} \ldots\ldots vn^{+} \right\}$$

$$a^{-} = \left\{ (v_{ij}, j \in J) \left(v_{ij} J \in J \right) J \right\} \tag{9}$$

$$v = \left\{ v1^{-}, v2^{-}, v3^{-} \ldots\ldots .vj^{-} \ldots\ldots vn^{-} \right\}$$

Here, $J = \{j = 1,2,3,\ldots\ldots .n\}$, $J' = \{j = 1,2,3,\ldots n\}$.

J and J' are associated with the beneficial and non-beneficial attributes.

Step 5: Here, the Euclidean distance of each alternative from the positive and negative ideal solution is calculated by using the following equations:

$$D_i^{+} = \sum_{i=1}^{n} \left(v_{ij} - v_i^{+} \right)^2, \ i = 1,2,3,\ldots,m \tag{10}$$

$$D_i^{-} = \sum_{i=1}^{n} \left(v_{ij} - v_i^{-} \right)^2, \ i = 1,2,3,\ldots,m \tag{11}$$

Step 6: Here, relative closeness to the ideal solution for each alternative is calculated by using the equation is given below:

$$C_i^{+} = \frac{D_i^{-}}{D_i^{+} + D_i^{-}}, \ i = 1,2,3,\ldots,m; 0 \le C_i^{+} \le 1 \tag{12}$$

Step 7: In the final step, ranking according to the preference order is given. The alternative with maximum relative closeness should be the best choice. $+C_i$ is multi-performance characteristic index (MPCI) in TOPSIS.

3. Experimental Procedure

The EDM at the Production Engineering Lab, Osmania University was used for coating. This machine is of CREATER make and numerical control (CNC) is shown in the process flow diagram. Firstly, titanium substrates were ground and polished with emery papers of 50, 100 and 200 micrometers. Then, the substrates were taken for preheat treatment at temperatures of 100 °C, 300 °C and 500 °C and quenched in brine, castor oil and vegetable oil in order to avoid workpiece erosion. Taguchi L9 was followed for heating temperatures as shown in Table 3. Preheat treatment was performed to increase the hardness that prevents workpiece erosion when coating. Before and after the heat treatment, hardness, grain size and grain area of each substrate were measured. Then, the substrates were taken for deposition following the input parameters shown in Table 3. Figure 1 depicts the steps followed in this work for coating copper on titanium alloy. Table 3 shows the MDR and EWR and Tables 4 and 5 shows the average hardness, average diameter, average grain area, average grain number and grain structure.

Table 3. Substrate Heating and Quenching and Experimental Data.

Exp. No.	Current (Amp)	Ton (µs)	Toff (µs)	Temp (°C)	Quenching Medium	MDR (Gram/Min)	EWR (Gram/Min)
1	4	280	200	100	Sunflower	0.00442	0.000442
2	4	360	300	300	Brine	0.00214	0.000852
3	4	440	400	500	Castor oil	0.005205	0.000682
4	8	280	300	500	Sunflower	0.002253	0.002775
5	8	360	400	100	Brine	0.003148	0.003565
6	8	440	200	300	Castor oil	0.00506	0.00462
7	12	280	400	300	Sunflower	0.012278	0.012423
8	12	360	200	500	Brine	0.001868	0.013923
9	12	440	300	100	Castor oil	0.000155	0.011508

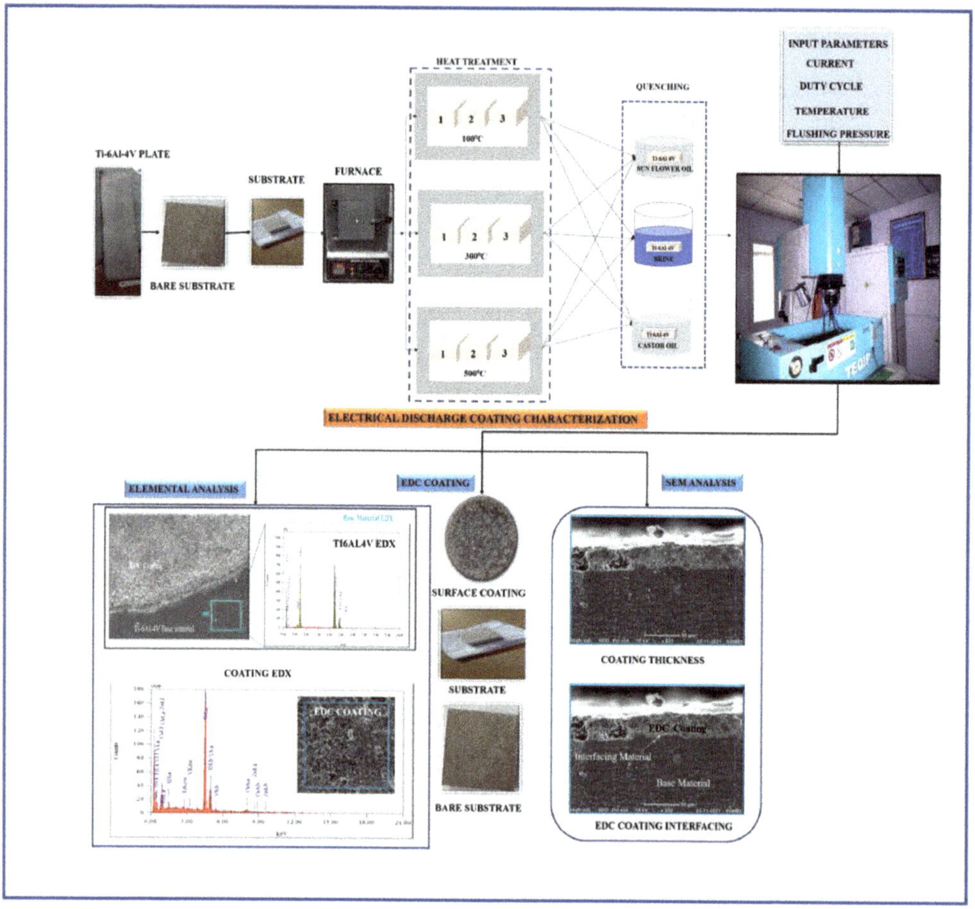

Figure 1. EDC Process Flowchart.

Table 4. Before Heating Hardness and Grain size.

Exp. No.	Avg. HV	Avg. Diameter (Micron)	Avg. Grain Area (Micron Sqr)	Avg. Grain No.	50X	Grain Structure
1	392.66	61.95	4865	4	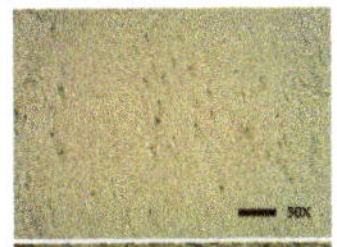	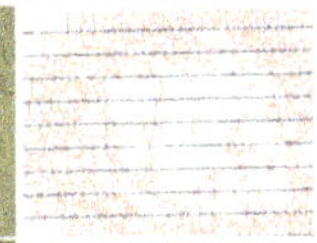
2	399.3	26.85	861.5	7		
3	350.46	26.85	861.5	7		
4	355.83	30.95	1220	6		
5	408.3	26.85	861.5	7		
6	389.5	30.95	1220	6		
7	393.56	30.95	1220	6		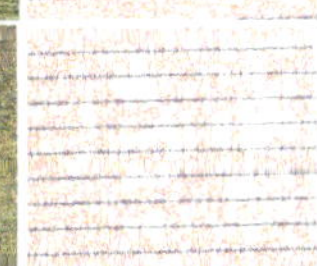
8	379.1	26.85	861.5	7		
9	378.96	36.8	1725	6		

Table 5. After Heating Hardness and Grain size.

Exp. No.	Avg. HV	Avg. Diameter (Micron)	Avg. Grain Area (Micron Sqr)	Avg. Grain No.	50X	Grain Structure
1	348.13	26.85	861.5	7		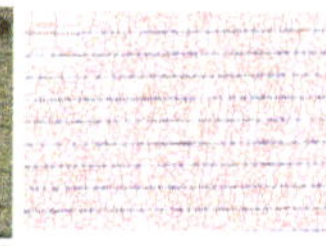
2	398.83	26.85	861.5	7		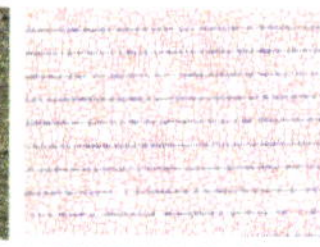
3	360.7	30.95	1220	6		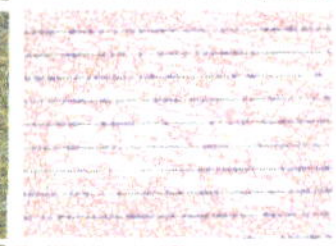
4	358.13	36.8	1725	6	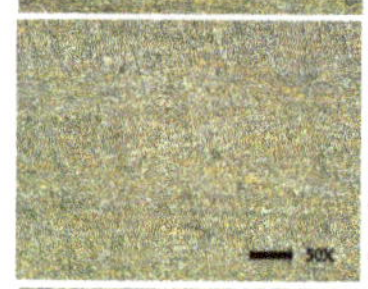	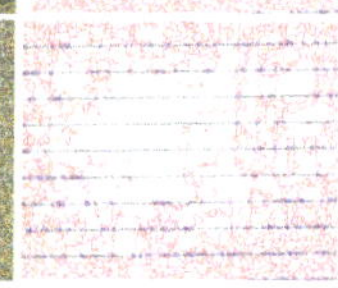
5	396.166	26.85	861.5	7		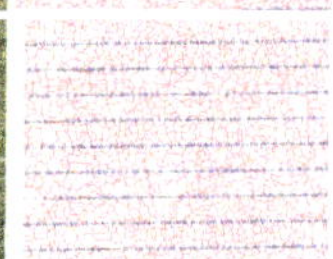
6	398.76	26.85	861.5	7		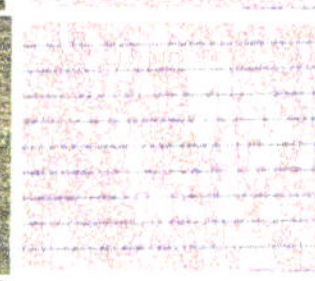
7	387.73	52.1	3440	5	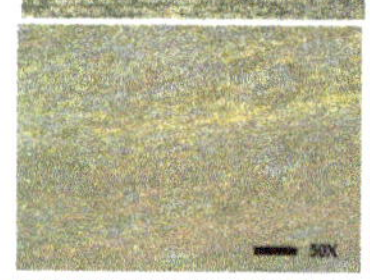	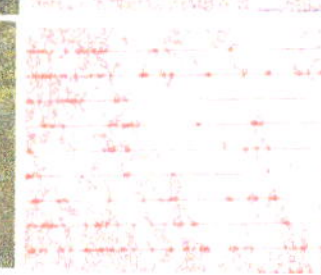
8	359.46	36.8	1725	6	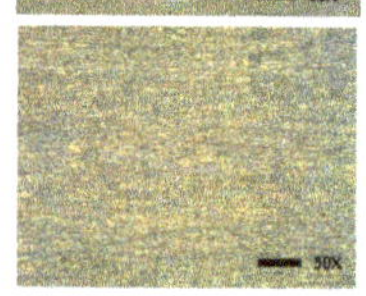	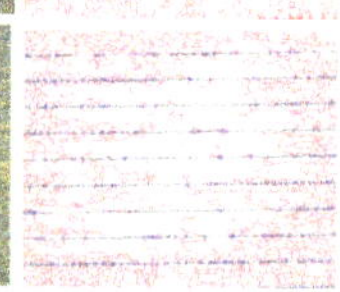
9	383.16	104.1	13750	3		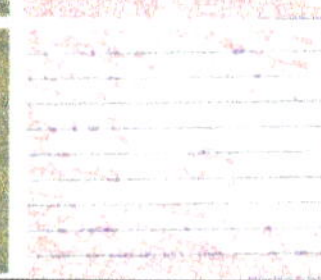

4. Results

This section covers the results obtained from the experiments and discussions on analysis of output parameters, surface integrity, surface characterization, surface crack density, coating interface analysis, elemental analysis and optimization by TOPSIS.

Analysis of Output Parameters

MDR, EWR, CT, elemental analysis and SCD are the output parameters considered in this study [37]. In EDC, output parameters are required to be studied since they depend on the input parameters and finding desired values for input parameters with respect to output parameters is difficult. Conditions to be followed for these parameters are: the higher the better for MDR, the lower the better for EWR and the lower the better for SCD [38]. From the literature, it was observed that lower values of EWR and SCD can be obtained by lower current, pulse-on time, pulse-off time and temperature [39]. According to the design of experiments, it is not possible to select the desired parameter values so there is a requirement of optimization techniques for this problem.

Table 6 shows the MDR, EWR, CT, SCD and elemental analysis obtained for all the experiments in which the maximum MDR and CT are 0.0122775 gram/min and 39.9 μm, respectively, whereas minimum values are 0.00044249 gram/min and 0.0000992775 μm^2, respectively. Figure 2 shows the graph of output parameters created with the table. Variation in output parameters with respect to variation in input parameters can be observed from Figure 2.

Table 6. Output response parameters.

Exp. No. Units	MDR (Gram/Min)	EWR (Gram/Min)	CT (μm)	SCD (μm^2)	Ti %	Al %	Cu %
1	0.00442	0.00044	38.3333	0.0000992775	86.45	1.45	7.65
2	0.00214	0.00085	28.8333	120971533.3	85.93	0.85	6.6
3	0.0052	0.00068	33.5667	156621668.1	90.21	1.39	4.72
4	0.00225	0.00278	27.5333	44156787.82	91.14	1.23	3.82
5	0.00315	0.00357	39.9	261218750.4	90.41	0.38	4.82
6	0.00506	0.00462	40.2	194348924	91.71	1.79	5.82
7	0.01228	0.01242	30.4333	188926795.8	91.57	1.41	5.24
8	0.00187	0.01392	39.8	117702032.4	92.42	1	2.11
9	0.00015	0.01151	31.7	121594295.4	91.18	1.23	3.84

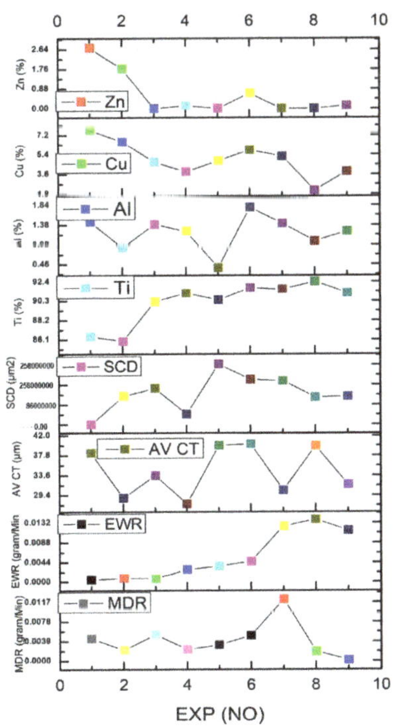

Figure 2. Experimental Graph.

5. Discussion

5.1. Analysis of Surface Integrity before and after Heat Treatment

For the study of surface integrity, all the substrates were measured for hardness individually and showed variation in average hardness ranging from 348 HV to 398 HV taken at six points. For the similar substrates, hardnesses were again measured after heat treatment. From Figure 3, it can be seen that experiments 3, 4, 6 and 9 have shown an increase in hardness after heat treatment at 100 °C, 300 °C, 300 °C and 500 °C, respectively, and quenched in castor oil, vegetable oil, castor oil and castor oil, respectively. An important observation among these was that all substrates quenched in brine solution have shown a decrease in hardness [40]. An increase of around 10 HV after heat treatment at 100 °C with quenching in castor oil was seen, so these parameters were selected.

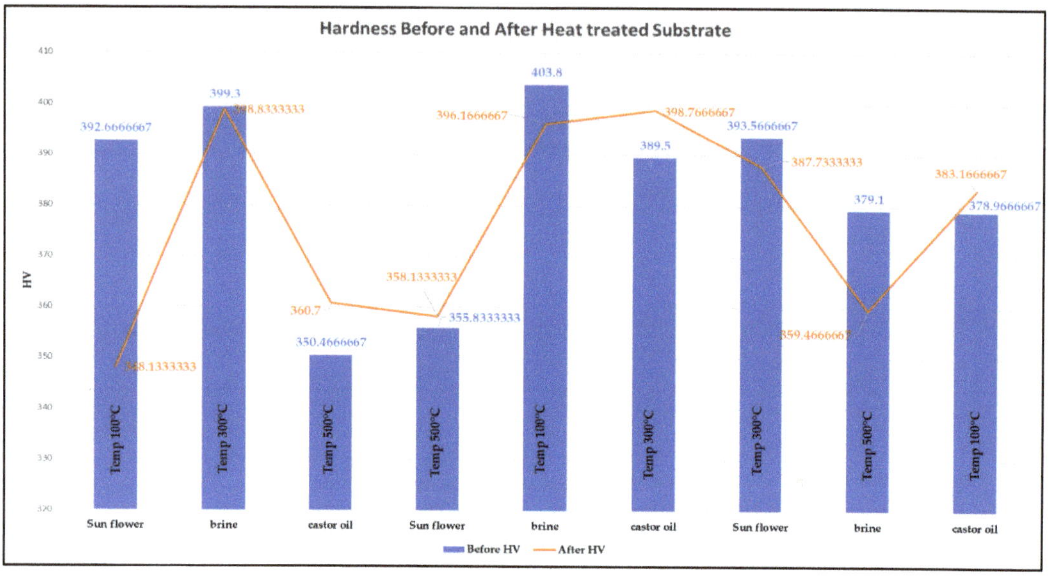

Figure 3. Hardness Before and After heat treatment of substrates.

5.2. Surface Morphology

Figure 4 shows the SEM images of EDCs developed using different combinations of input parameters with the L9 orthogonal array. While designing the experiments, the process of heating treatment and quenching medium were also considered. It is observed from Figure 4 that all the coatings have a cauliflower structure and uniform coating. Figure 4b,d,g show uneven coating surfaces and machining spatters can be observed. Figure 7c shows uniform coating on the substrate at 100 °C with quenching in castor oil. Input parameters for experiment 3 are current of 4 Amp, Ton of 440 µs and Toff of 400 µs.

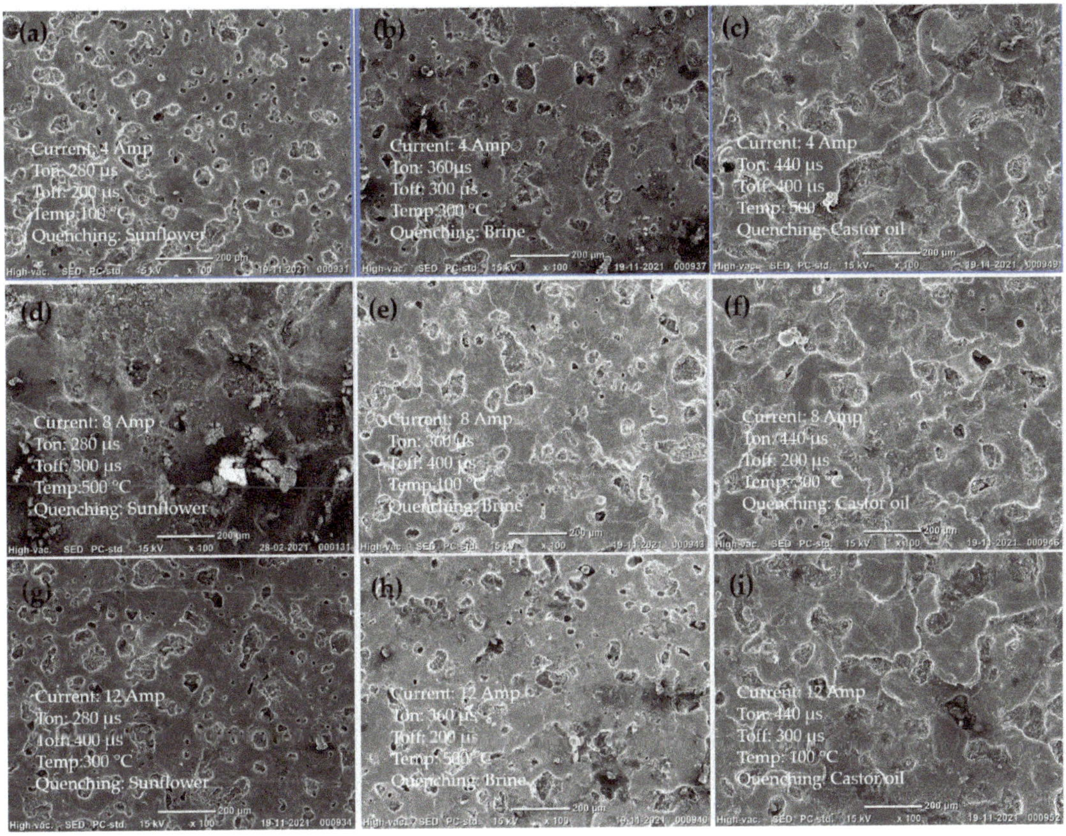

Figure 4. EDC coating Surface observations: (**a**) EX 1, (**b**) EX 2, (**c**) EXP 3, (**d**) EX 4, (**e**) EX 5 and (**f**) EX 6, (**g**) EX 7, (**h**) EX 8 and (**i**) EX 9 (all SEM images are 100X).

5.3. Surface Crack Density

Figure 5 shows the surfaces of coatings captured using the scanning electron microscope. Coatings were thoroughly examined using SEM and if a crack was found, it was zoomed into with a magnification of 500X. Crack length was measured with SEM and SCD was calculated for all the coatings as per Equation (3) [41]. Cracks were observed in almost all the coatings and a minimum crack density of 0.000099277 µm/µm^2 was ob-tained for Figure 5a.

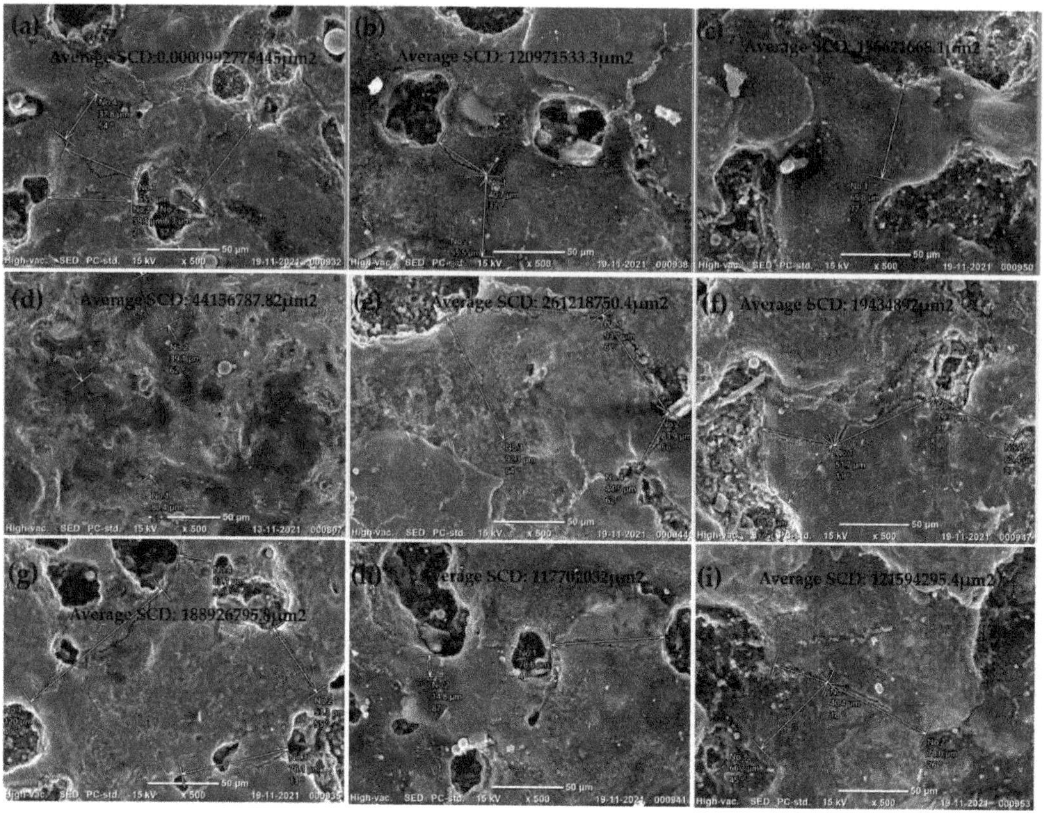

Figure 5. EDC coating Surface with observed microcracks: (**a**) EX 1, (**b**) EX 2, (**c**) EXP 3, (**d**) EX 4, (**e**) EX 5 and (**f**) EX 6, (**g**) EX 7, (**h**) EX 8 and (**i**) EX 9 (all SEM images are 500X).

5.4. Coating and Base Material Interfacing Analysis

Figure 6 depicts the interfacing and bonding of coating on the base material. The investigation of copper coatings obtained on preheated substrates showed major variations in the CT which is also a function of process conditions. SEM images of the cross section of coatings deposited under different conditions are shown in Figure 6. It was observed that with an increase in current, heat is generated and damages the base material, as shown in Figure 6e,g,i. For these experiments, the deposition rate was high due to the current and duty cycle. The highest CT of 40.2 μm can be observed from Figure 6f but it seems to be highly discontinuous with the parameter combination of current of 8 Amp, Ton of 440 μs, Toff of 200 μs, preheated temperature of 300 °C and quenching in castor oil. Figure 7 shows the graph with CTs along with MDR and EWR. From this figure, it can be observed that higher CT does not necessarily mean high MDR and EWR.

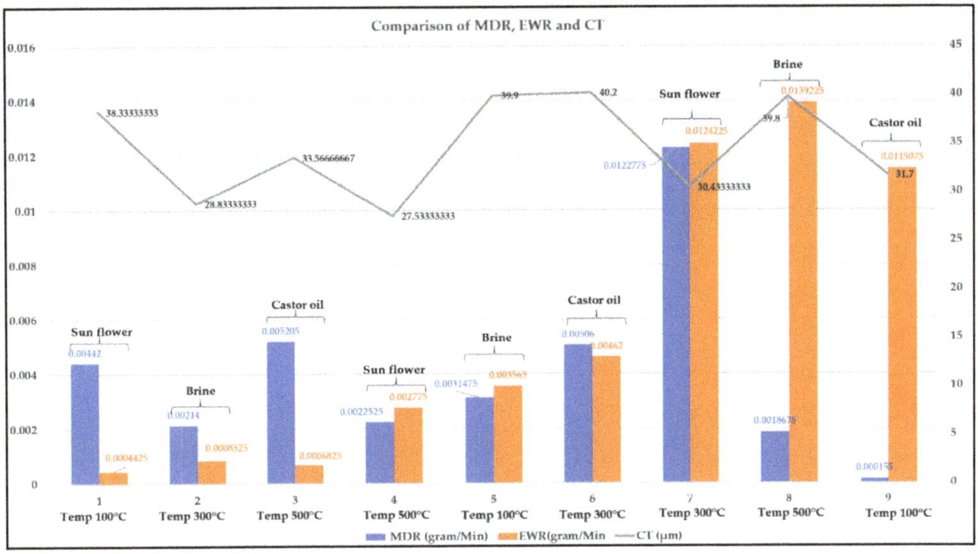

Figure 6. Comparison of MDR, EWR and CT.

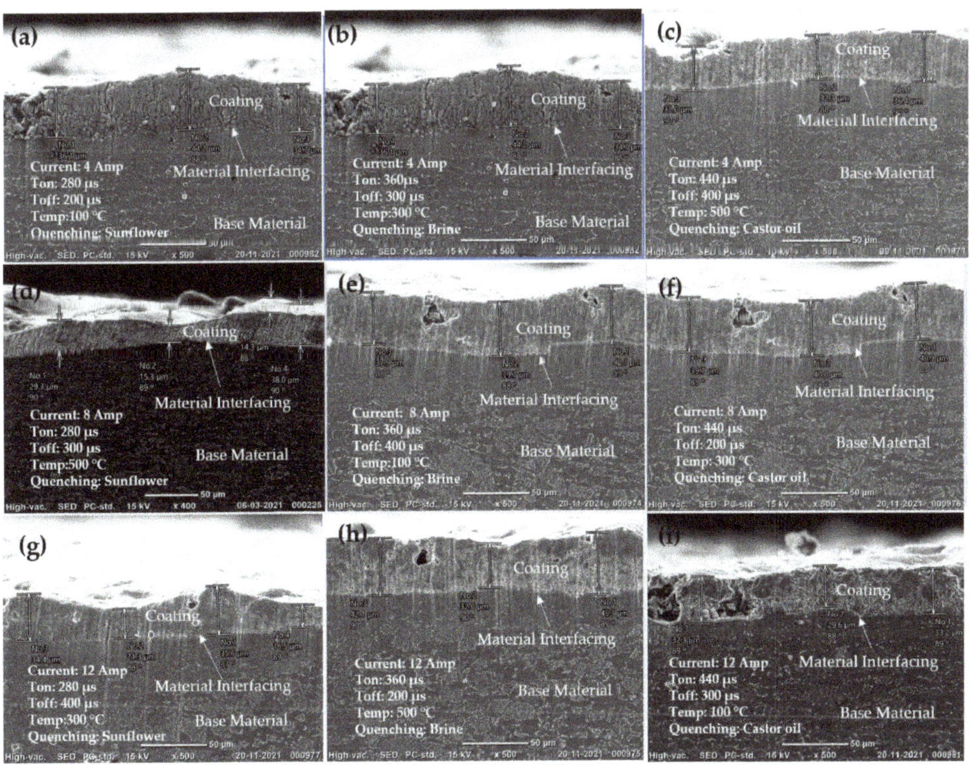

Figure 7. EDC interaction between the base material and coating: (**a**) EX 1, (**b**) EX 2, (**c**) EXP 3, (**d**) EX 4, (**e**) EX 5 and (**f**) EX 6, (**g**) EX 7, (**h**) EX 8 and (**i**) EX 9 (all SEM images are 500X).

5.5. Elemental Analysis

SEM was used to inspect the composition of obtained coatings with energy dispersive X-ray spectroscopy (EDX). It was understood from Figure 8 that a higher Ti and Cu percentage was obtained in the coating deposited with the parameters of experiment 8 (Figure 8h), being Ti 92.42%, current 12 Amp, Ton 360 µs, Toff 200 µs and temperature 500 °C and with quenching in brine solution. Meanwhile, the substrate that had the highest copper percentage (7.65%) was coated with the input parameters of current 4 Amp, Ton 280 µs, Toff 200 µs, temperature 100 °C and quenching in sunflower oil.

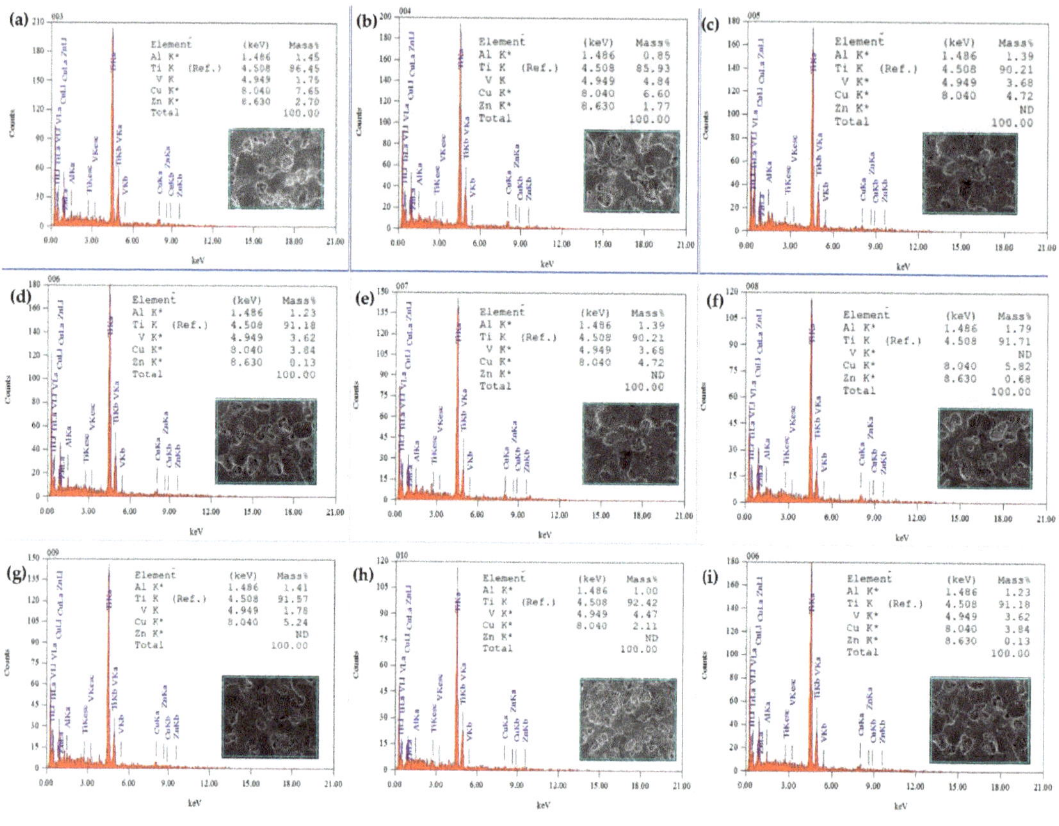

Figure 8. EDC Coating Surface EDX: (**a**) EX 1, (**b**) EX 2, (**c**) EXP 3, (**d**) EX 4, (**e**) EX 5 and (**f**) EX 6, (**g**) EX 7, (**h**) EX 8 and (**i**) EX 9.

5.6. Optimization by TOPSIS

From the discussion, it can be understood that an optimization technique is required to select the optimum coating among all these coatings. It is difficult to select one manually because each experiment was best with some output parameter. So, the TOPSIS optimization technique was applied to select the optimum coating [42]. TOPSIS is an optimization technique involving seven steps [4]. The formula used to calculate at each step was described in the Section 2. The first step is to form a matrix using the output parameters that support simplifying and processing easily and efficiently, as shown Table 7. Then, further steps were followed as per Section 4.

Table 7. Decision matrix.

Exp. No.	MDR	EWR	AV CT	SCD	Ti	Cu
1	0.00442	0.000442	38.33333	0.0000993	86.45	7.65
2	0.00214	0.000852	28.83333	121000000	85.93	6.6
3	0.005205	0.000682	33.56667	157000000	90.21	4.72
4	0.002252	0.002775	27.53333	44156788	91.14	3.82
5	0.003147	0.003565	39.9	261000000	90.41	4.82
6	0.00506	0.00462	40.2	194000000	91.71	5.82
7	0.012278	0.012423	30.43333	189000000	91.57	5.24
8	0.001868	0.013923	39.8	118000000	92.42	2.11
9	0.000155	0.011508	31.7	122000000	91.18	3.84

The last step is to calculate the relative closeness using the formulae shown in Equation (12). The values are tabulated in Table 8, and it can be observed that experiment 6 has higher closeness and takes the rank of 1. This represents the coating obtained from experiment 6 which is the optimum coating as per the conditions of the higher the better MDR and the lower the better EWR and SCD.

Table 8. Relative closeness.

Exp. No.	Relative Closeness	Rank
1	0.370044	3
2	0.370065	9
3	0.370053	5
4	0.370064	8
5	0.370041	2
6	0.370034	1
7	0.370054	6
8	0.370046	4
9	0.370057	7

6. Conclusions

In this work, a bronze electrode was selected to coat copper on titanium alloy (Ti6Al4V) to compare with our previous attempts. Prior to coating, workpiece substrates were preheated at different temperatures of 100 °C, 300 °C and 500 °C and quenched in brine, castor oil and vegetable oil in order to avoid workpiece erosion. After this treatment, hardness, grain area, grain diameter and number of grains were characterized to compare with pretreated substrates. EDC input parameters selected to be optimized were current, Ton, Toff and preheating temperature. The TOPSIS technique was used to optimize the input parameters and material characterization was conducted using SEM with EDX. Some of the conclusions from this study are as follows:

Experiments were carried out according to the Taguchi L9 design of experiments. A higher increase (10 HV) in hardness was obtained for substrate heat treated at 100 °C and quenched in castor oil. It was observed that MDR increases with a decrease in current and EWR increases with an increase in current and Ton.

Surface morphology of all coatings showed a cauliflower structure.

SEM with EDX confirmed a maximum copper percentage of 7.65% in the coating surface whereas copper coated with brass electrodes in our previous study had up to 70% copper material when experiments were performed with the same experimental conditions. The highest coating thickness of 40.2 μm was obtained for experiment 6 when observed in SEM images of magnification of 500X.

Finally, TOPSIS has ranked experiment number six with the input process parameters of current 8 Amp, Ton 440 μs, Toff 200 μs, temperature 300 °C and quenching medium of castor oil as the optimum. Output response values of the same experiments are MDR

0.00506 gram/min, EWR 0.00462 gram/min, CT 40.2 μm and SCD 194348924.μm^2, with Ti 91.71%, Cu 5.82%, Zn 0.68% and hardness 398.767 HV.

Author Contributions: Conceptualization, J.M. and R.U.S.; methodology, J.M. and R.U.S.; software, J.M. and R.U.S.; validation, R.U.S. and J.M.; formal analysis, J.M.; investigation, J.M.; resources, B.K.; data curation, J.M.; writing—original draft preparation, J.M. and R.U.S.; writing—review and editing, R.U.S.; visualization, J.M. and R.U.S.; Supervision, B.K., H.E. and K.A.; project administration, J.M. and R.U.S.; funding acquisition, B.K. All authors have read and agreed to the published version of the manuscript.

Funding: This research received no external funding.

Data Availability Statement: Not applicable.

Acknowledgments: The authors would like to thank S Arun Kumar, Technician, Production Lab; A Krishnaiah, Production Lab In-Charge and Rega Rajendra, Head, Department of Mechanical Engineering, University College of Engineering, Osmania University, for their support during this work.

Conflicts of Interest: The authors declare no conflict of interest.

References

1. Banu, A.; Ali, M.Y. Ali, Electrical Discharge Machining (EDM): A Review Electrical Discharge Machining (EDM): A Review. *Int. J. Eng. Mater. Manuf.* **2016**, *1*, 3–10. [CrossRef]
2. Faisal, N.; Kumar, K. Optimization of Machine Process Parameters in EDM for EN 31 Using Evolutionary Optimization Techniques. *Technologies* **2018**, *6*, 54. [CrossRef]
3. Shabgard, M.R.; Gholipoor, A.; Baseri, H. A review on recent developments in machining methods based on electrical discharge phenomena. *Int. J. Adv. Manuf. Technol.* **2016**, *87*, 2081–2097. [CrossRef]
4. Descoeudres, A. Characterization of Electrical Discharge Machining Plasmas. *EPFL* **2006**, *3542*, 137. [CrossRef]
5. Liew, P.J.; Yap, C.Y.; Wang, J.; Zhou, T.; Yan, J. Surface modification and functionalization by electrical discharge coating: A comprehensive review. *Int. J. Extreme Manuf.* **2020**, *2*, 012004. [CrossRef]
6. Maddu, J.; Shaik, R.U. A Review on Electrical Discharge Coating (EDC) and its Multi-Optimization Techniques. *IOP Conf. Ser. Mater. Sci. Eng.* **2021**, *1185*, 12027. [CrossRef]
7. Murray, J.; Algodi, S.; Fay, M.; Brown, P.; Clare, A. Formation mechanism of electrical discharge TiC-Fe composite coatings. *J. Mater. Process. Technol.* **2017**, *243*, 143–151. [CrossRef]
8. Krishna, M.E.; Patowari, P.K. Parametric Study of Electric Discharge Coating using Powder Metallurgical Green Compact Electrodes. *Mater. Manuf. Process.* **2014**, *29*, 1131–1138. [CrossRef]
9. Mughal, M.P.; Farooq, M.U.; Mumtaz, J.; Mia, M.; Shareef, M.; Javed, M.; Jamil, M.; Pruncu, C.I. Surface modification for osseointegration of Ti6Al4V ELI using powder mixed sinking EDM. *J. Mech. Behav. Biomed. Mater.* **2020**, *113*, 104145. [CrossRef]
10. Machkale, P.; Dabade, B. Experimental investigation of tungsten and copper carbide coating on AISI 1020 steel using electro discharge coating process. *Mater. Today Proc.* **2019**, *26*, 2915–2920. [CrossRef]
11. Mohanty, S.; Bhushan, B.; Das, A.K.; Dixit, A.R. A study on parametric optimization of Micro-electrical discharge coating process using response surface methodology. *Mater. Today Proc.* **2020**, *38*, 325–332. [CrossRef]
12. Maddu, J.; Karrolla, B.; Shaik, R.U.; Vuppala, S. Comparative Study of Optimization Models for Evaluation of EDM Process Parameters on Ti-6Al-4V. *Modelling* **2021**, *2*, 29. [CrossRef]
13. Sumanth, P.; Reddy, M.V.; Shaik, I.; Maddu, J.R. Parameter Optimization of EDM Characteristics on Ti-6Al-4V Using Different Electrodes. *IOP Conf. Ser. Mater. Sci. Eng.* **2021**, *1185*, 012022. [CrossRef]
14. Norambuena, G.A.; Patel, R.; Karau, M.; Wyles, C.C.; Jannetto, P.J.; Bennet, K.E.; Hanssen, A.D.; Sierra, R.J. Antibacterial and Biocompatible Titanium-Copper Oxide Coating May Be a Potential Strategy to Reduce Periprosthetic Infection: An In Vitro Study. *Clin. Orthop. Relat. Res.* **2017**, *475*, 722–732. [CrossRef]
15. Tyagi, R.; Darmalingam, K.; Patel, V.S.; Das, A.K.; Mandal, A. Mandal, Deposition of WS2 and Cu nanopowder coating using EDC process and its analysis. *Mater. Today Proc.* **2019**, *18*, 5170–5176. [CrossRef]
16. Algodi, S.J.; Murray, J.W.; Brown, P.D.; Clare, A.T. Wear performance of TiC/Fe cermet electrical discharge coatings. *Wear* **2018**, *402–403*, 109–123. [CrossRef]
17. Mussada, E.K.; Patowari, P.K. Post processing of the layer deposited by electric discharge coating. *Mater. Manuf. Process.* **2017**, *32*, 442–449. [CrossRef]
18. Hsu, H.-Y.; Hu, C.-C. Surface quality improvement of EDMed Ti–6Al–4V alloy using plasma etching and TiN coating. *Int. J. Adv. Manuf. Technol.* **2017**, *88*, 67–74. [CrossRef]
19. Maddu, J.; Vuppala, S.; Shaik, R.U. Formation and Statistical Optimization of Electrical Discharge Coatings using Conventional Electrodes. *Energies* **2021**, *14*, 5691. [CrossRef]
20. Algodi, S.J.; Murray, J.W.; Fay, M.W.; Clare, A.T.; Brown, P.D. Electrical discharge coating of nanostructured TiC-Fe cermets on 304 stainless steel. *Surf. Coatings Technol.* **2016**, *307*, 639–649. [CrossRef]

21. Tyagi, R.; Das, A.; Mandal, A. Electrical discharge coating using WS2 and Cu powder mixture for solid lubrication and enhanced tribological performance. *Tribol. Int.* **2018**, *120*, 80–92. [CrossRef]
22. Murray, J.; Cook, R.; Senin, N.; Algodi, S.; Clare, A. Defect-free TiC/Si multi-layer electrical discharge coatings. *Mater. Des.* **2018**, *155*, 352–365. [CrossRef]
23. Bui, V.D.; Mwangi, J.W.; Meinshausen, A.-K.; Mueller, A.J.; Bertrand, J.; Schubert, A. Antibacterial coating of Ti-6Al-4V surfaces using silver nano-powder mixed electrical discharge machining. *Surf. Coat. Technol.* **2020**, *383*, 125254. [CrossRef]
24. Ming, W.; Guo, X.; Xu, Y.; Zhang, G.; Jiang, Z.; Li, Y.; Li, X. Progress in non-traditional machining of amorphous alloys. *Ceram. Int.* **2023**, *49*, 1585–1604. [CrossRef]
25. Kahlin, M.; Ansell, H.; Basu, D.; Kerwin, A.; Newton, L.; Smith, B.; Moverare, J.J. Improved fatigue strength of additively manufactured Ti6Al4V by surface post processing. *Int. J. Fatigue* **2020**, *134*, 105497. [CrossRef]
26. Zhang, Z.; Liu, D.; Zhang, Y.; Xue, T.; Huang, Y.; Zhang, G. Fabrication and droplet impact performance of superhydrophobic Ti6Al4V surface by laser induced plasma micro-machining. *Appl. Surf. Sci.* **2022**, *605*, 154661. [CrossRef]
27. Schnell, G.; Duenow, U.; Seitz, H. Effect of Laser Pulse Overlap and Scanning Line Overlap on Femtosecond Laser-Structured Ti6Al4V Surfaces. *Materials* **2020**, *13*, 969. [CrossRef]
28. Zeng, Z.-Y.; Xiao, H.-Q.; Jie, X.-H.; Zhang, Y.-M. Friction and wear behaviors of TiCN coating based on electrical discharge coating. *Trans. Nonferrous Met. Soc. China* **2015**, *25*, 3716–3722. [CrossRef]
29. Jeavudeen, S.; Jailani, H.S.; Murugan, M. Effect of process parameters in the machining of Titanium alloy and high speed steel in powder mixed electrical discharge machining process. *Mater. Today Proc.* **2020**, *27*, 615–619. [CrossRef]
30. Maddu, J.; Karrolla, B.; Shaik, R.U. Experimental optimization of electrical discharge coatings using conventional electrode. *Mater. Sci. Eng. B* **2022**, *286*, 116069. [CrossRef]
31. Kumar, N.; Mandal, N.; Das, A.K. Micro-machining through electrochemical discharge processes: A review. *Mater. Manuf. Process.* **2020**, *35*, 363–404. [CrossRef]
32. Jagadeeswara Rao, M.; Shaik, R.U.; Buschaiah, K. Electrical Discharge Machining: A Comparative Surface Integrity Study for Incoloy-800. *Mater. Today Proc.* **2020**, *22*, 3286–3296. [CrossRef]
33. Mansor, A.F.; Azmi, A.I.; Zain, M.Z.; Jamaluddin, R. Parametric evaluation of electrical discharge coatings on nickel-titanium shape memory alloy in deionized water. *Heliyon* **2020**, *6*, e04812. [CrossRef] [PubMed]
34. ASahu, A.K.; Thomas, J.; Mahapatra, S.S. An intelligent approach to optimize the electrical discharge machining of titanium alloy by simple optimization algorithm. *Proc. Inst. Mech. Eng. Part E J. Process. Mech. Eng.* **2021**, *235*, 371–383. [CrossRef]
35. Buschaiah, K.; JagadeeswaraRao, M.; Krishnaiah, A. Krishnaiah, Investigation on the Influence of Edm Parameters on Machining Characteristics for Aisi 304. *Mater. Today Proc.* **2018**, *5*, 3648–3656. [CrossRef]
36. Mohanty, S.; Kumar, V.; Das, A.K.; Dixit, A.R. Dixit, Surface modification of Ti-alloy by micro-electrical discharge process using tungsten disulphide powder suspension. *J. Manuf. Process.* **2018**, *37*, 28–41. [CrossRef]
37. Kumar, D.; Mondal, S. Process parameters optimization of AISI M2 steel in EDM using Taguchi based TOPSIS and GRA. *Mater. Today Proc.* **2019**, *26*, 2477–2481. [CrossRef]
38. Meel, R.; Singh, V.; Katyal, P.; Gupta, M. Optimization of process parameters of micro-EDD/EDM for magnesium alloy using Taguchi based GRA and TOPSIS method. *Mater. Today Proc.* **2021**, *51*, 269–275. [CrossRef]
39. Izwan, N.S.L.B.; Feng, Z.; Patel, J.B.; Hung, W.N. Prediction of Material Removal Rate in Die-sinking Electrical Discharge Machining. *Procedia Manuf.* **2020**, *5*, 658–668. [CrossRef]
40. Shaik, R.U.; Karanjai, M.; Joardar, J.; Hebalkar, N.Y.; Borse, P.H. Thermally stable electro catalytic nickel-phosphide film deposition on graphite for HER electrode application. *Mater. Sci. Eng. B* **2022**, *285*, 115927. [CrossRef]
41. Huang, C.H.; Yang, A.B.; Hsu, C.Y. The optimization of micro EDM milling of Ti–6Al–4V using a grey Taguchi method and its improvement by electrode coating. *Int. J. Adv. Manuf. Technol.* **2018**, *96*, 3851–3859. [CrossRef]
42. Maddu, J.; Karrolla, B.; Shaik, R.U.; Burdhuhos-Nergis, D.-P. SWOT Analysis of Electrical Discharge Coatings: A Case Study of Copper Coating on Titanium Alloy. *Surfaces* **2022**, *5*, 21. [CrossRef]

micromachines

Article

Optimization of Fatigue Performance of FDM ABS and Nylon Printed Parts

Andrey Yankin, Gaini Serik, Saniya Danenova, Yerassyl Alipov, Ali Temirgali, Didier Talamona and Asma Perveen *

Department of Mechanical and Aerospace Engineering, School of Engineering and Digital Sciences, Nazarbayev University, Astana 010000, Kazakhstan
* Correspondence: asma.perveen@nu.edu.kz; Tel.: +7-(7172)-70-91-32

Abstract: This research work aims to proceed with the optimization of Fused Deposition Modeling (FDM) printing parameters for acrylonitrile butadiene styrene (ABS) and polyamide (Nylon) to improve fatigue resistance. For that purpose, the methodology of the paper involves two main approaches: experimental study and finite element analysis. The experimental part of the paper used the Taguchi method to find the effects of printing internal geometry, printing speed, and nozzle diameter on the fatigue life of ABS and Nylon plastic materials. ANCOVA multiple linear regression and sensitivity analysis was used to investigate the effects of printing parameters on the fatigue life of materials. The analysis of the results revealed: Nylon performed better than ABS, but had a higher slope; the 'tri-hexagon' structure resulted in the highest fatigue life, but the effect was statistically significant only for ABS material; the fatigue life of both materials increased with decreasing the nozzle diameter; the printing speed had no statistically significant influence neither on ABS nor Nylon. The experimental results then were validated by numerical simulations and the difference between the values was within ±14% depending on the experiment. Such differences might occur due to numerical and experimental errors.

Keywords: Fused Deposition Modeling (FDM); acrylonitrile butadiene styrene (ABS); polyamide (Nylon); fatigue; ANCOVA regression analysis; Taguchi analysis; parametric study; numerical study

Citation: Yankin, A.; Serik, G.; Danenova, S.; Alipov, Y.; Temirgali, A.; Talamona, D.; Perveen, A. Optimization of Fatigue Performance of FDM ABS and Nylon Printed Parts. *Micromachines* **2023**, *14*, 304. https://doi.org/10.3390/mi14020304

Academic Editor: Nikolaos Tapoglou

Received: 24 November 2022
Revised: 15 January 2023
Accepted: 17 January 2023
Published: 24 January 2023

1. Introduction

There are a variety of different additive manufacturing techniques, one that the object of interest here is fused deposition manufacturing known as FDM. It involves a nozzle that contains molted filament to transform on a 2D plane plate to build up a layer of a cross-sectional region of a digitalized object. Then the platform moves in up or down directions to continue the layer-by-layer construction process [1].

The FDM process involves thermoplastic materials of high strength such as polycarbonate (PC), polylactic acid (PLA), acrylonitrile butadiene styrene (ABS), polyamide (Nylon), etc. [2–5]. The application of FDM printing is essential when it comes to inexpensive parts in a short period of time, rigid models, and the construction of a prototype for validation purposes. Thus, the areas of application of 3D printed parts using FDM technology and thermoplastic materials are rapidly increasing and currently involve automotive, aerospace, medical, industrial, manufacturing, architecture, etc. [6,7].

In spite of the fact that there are many advantages of FDM printing over conventional methods, the limitations of this technique are varied in range from low-lying mechanical properties to lousy surface quality of printed parts. One of the most significant challenges is associated with voids formation between layers of printed parts [8]. This occurs owing to the weak interaction between the extruded layer and solidified part [9]. Therefore, it is crucial to examine the mechanical properties of finalized printed objects.

The most common process parameters are infill density, infill patterns (internal geometric structure), extrusion temperature, nozzle diameter, layer thickness, raster angle, build orientation printing speed, etc. [5,10,11]. Based on the literature review findings, infill density tends to be one of the primary factors affecting mechanical behavior. For example, the strength properties of the printed ABS, PLA, and Nylon increase as infill density increases [12–15]. By so doing, this parameter can be varied from 0% (hollow part) to 100% (solid part).

Along with infill density, the internal geometric structure plays a crucial role in the mechanical properties. It defines the interaction between infilled filaments at the moment of being loaded. Some of the commonly used infill patterns are hexagonal, linear, and diamond. It should be noted that one pattern could provide better results for tensile or compressive properties while the same pattern may not hold well for a component subjected to other types of load [16].

Holding under the control of the relationship between nozzle size and layer thickness helps to regulate the air gaps in the structural matrix of the FDM printed part [16]. Increasing the nozzle diameter relative to layer thickness results in higher flexural strength. This is because increasing the ratio of the nozzle to layer thickness produces more contact surfaces between the layers of printed parts [17].

The printing speed parameter is associated with how fast the nozzle and the rest of the transformable parts of a 3D printer move relatively to stationary elements. The target idea of this parameter is to compromise the duration of printing and the quality of printed parts. At too high a printing speed, the model ends up with a weak interaction between extruded layer and solidified part. Moreover, there are still plenty of parameters that could be varied during the manufacturing process. More information about them may be found in some review papers [5,10,18,19].

FDM components are often subjected to cyclic loads, which might be a reason for fatigue failure. It is evident that the fatigue properties also depend on process parameters. For instance, it was illustrated that fill density, nozzle diameter, and layer thickness were the most influential factors in the fatigue life of PLA [20,21]. The optimal filling pattern for the Nylon was investigated in the paper [22]. As a result, specimens with a triangular filling pattern and matrix density of 20%, reinforced with carbon fiber at 0 degrees, showed better fatigue performance.

The effects of the print direction for PLA [23,24], ABS [23,25], and Nylon [26] as well as raster orientation for PLA [27] and ABS [28] components were explored. These works reveal that print direction and raster orientation have a significant effect on the fatigue performance of FDM materials. In addition, Ziemian et al. [29] investigated the characteristics of fatigue damage accumulation using FDM specimens for multiple mesostructure combinations, i.e., fiber orientations and layering patterns.

Some compare studies are presented in articles [30–32]. Namely, Miller et al. [32] conducted a comparative analysis of the fatigue performance of three PCU materials with systematically varied hard and soft segment contents, processed by both injection molding and FDM. In [30,31], Terekhina et al. compared the fatigue performance of various polyamides produced by FDM and selective laser sintering.

In the article [33], the authors applied a slightly different approach. At first, they conducted monotonic torsion tests at various infill patterns to obtain the best configuration for ABS. After that, the best parameter configuration was chosen to examine the fatigue properties at fully reversed load-controlled loading conditions. In [34], authors determined the S-N curve constants for ABS at fixed printing parameters. In the end, it is worth noticing some paper reviews of the fatigue behavior of 3D-printed polymers [23,35]. The literature review summary is presented in Table 1.

Table 1. The literature review summary.

Material	Printed Parameters	Results	Refs.
ABS	- nozzle temperature: 220–230 °C - nozzle diameter: 0.5 mm - printing speed: 30 mm/s - layer height: 0.1 mm - infill density: 100%	The average cycle number under the load of 30 N is 3796 cycles. The average cycle number is decreased to 128 cycles when the load is 60 N.	[34]
ABS	- melting temperature: 220 °C - nozzle diameter: 0.5 mm - printing speed: 40 mm/s - layer height: 0.15 mm - infill density: 70%	The parameters for Basquin's equation that constitutes S-N curve were found with A = 63.31 MPa, and b = −0.204.	[33]
ABS	- nozzle temperature: 245 °C - infill density: 50% - infill pattern: square - printing speed: 60 mm/s - nozzle diameter: 0.4 mm - layer height: 0.15 mm - print direction: horizontal and vertical	The parameters for the Basquin's equation that constitutes SN curve were found with A = 167.26 MPa, and b = −0.2782 (Horizontal); A = 395.67 MPa, and b = −0.3831 (Vertical)	[23]
ABS	- road width 0.3048 mm - slice height 0.1778 mm - part interior fill style: Solid normal - part fill style: Perimeter/raster - liquefier temperature: 320 °C	The specimens with raster orientation of +45/−45° had the longest fatigue life at each normalized stress level, followed by the 0, 45 and 90° orientations in descending order.	[28]
Nylon	- melting temperature: 195 °C - raster orientation: 0 deg - printing speed: 37.5 mm/s - layer height: 0.2 mm - nozzle diameter: 0.5 mm - number of contours: 3 - filling percentage: 100%	XZ build orientation of Nylon reveals a higher overall fatigue life than that for the XY one	[26]
Nylon CF	- melting temperature: 210 °C - nozzle diameter: 0.3 mm - printing speed: 30 mm/s - layer height: 0.1 mm - infill density: 50% - infill pattern: triangular, hexagonal - filling percentage (%): 20, 50 - filling layers: 24 - walls: 2	The parameters for the Basquin's equation that constitutes S-N curve were found with A = 206 MPa, and b = −0.039 for specimens with a triangular filling pattern and matrix density of 20%, reinforced with carbon fiber at 0 degrees.	[22]
Nylon (PA12, PA6)	- melting temperature: 250, 195 °C - raster orientation: 0 deg - printing speed: 32, 37.5 mm/s - layer height: 0.2 mm - nozzle diameter: 0.5 mm - number of contours: 3 - filling percentage: 100%	PA12 has a higher overall fatigue life than PA6. The fatigue behavior of PA12 FDM specimens is more resistant than the one obtained by SLS.	[30,31]

Though there are scholarly works that focus on examining the fatigue behavior of FDM printed parts, research works are mainly focused on determining the S-N curve for materials and provide vague studies on parametric analysis. As a result, the improvement in fatigue performance through optimization of printing parameters has a poor reflection in scholarly papers. Furthermore, there is much less fatigue research on Nylon compared to other plastics, for example, ABS. Thus, this study is focused on the determination of the optimization factor conditions to improve the fatigue properties of two thermoplastics ABS and Nylon produced by FDM.

2. Materials and Methods

This study is focused on two types of thermoplastics for FDM printing such as ABS and PA6 (Nylon). According to the literature review, three parameters were chosen to study the fatigue behavior of the FDM printed parts: nozzle diameter, internal geometric structure, and printing speed. Table 2 indicates the parameters and their levels.

Table 2. Selected independent variables for fatigue tests.

Parameter	L1	L2	L3
Nozzle diameter	0.2 mm	0.4 mm	0.6 mm
Internal geometry	Tri-hexagonal	Zigzag	Concentric
Printing speed	25 mm/s	30 mm/s	35 mm/s

The design of experiments was conducted within the Taguchi method to identify significant parameters and their levels for upgrading the fatigue behavior of the 3D printed part by performing a number of experiments. This approach allows a simple and inexpensive method in the various branches involving minimum experimental runs [36]. Such techniques enable researchers and engineers to decrease the utilization of material used, energy consumption, and environmental pollution impact. The Taguchi L9 Orthogonal Array with three levels for each three selected parameters is shown in Table 3.

Table 3. L_9 Orthogonal Array.

Test Code	Nozzle Diameter	Internal Geometry	Printing Speed
111	1	1	1
122	1	2	2
133	1	3	3
213	2	1	3
222	2	2	2
231	2	3	1
313	3	1	3
321	3	2	1
332	3	3	2

The parameters for printing were modified using Ultimaker Cura software 4.3 and an 'STL' input file with a designated 3D part. Apart from the parameters that were examined, there are constant variables that are summarized in Table 4. The extrusion temperature for ABS was 220 °C and 210 °C for Nylon. The building orientation and internal geometry of a specimen are presented in Figure 1.

Table 4. Constant variables for printing.

Parameter	Value
Layer Height	0.15 mm
Orientation	horizontal
Wall Thickness	1.3 mm
Wall Line Count	4
Horizontal Expansion	0 mm
Top/Bottom thickness	1.2 mm
Top Layers	8
Bottom Layers	8
Fan Speed	2%

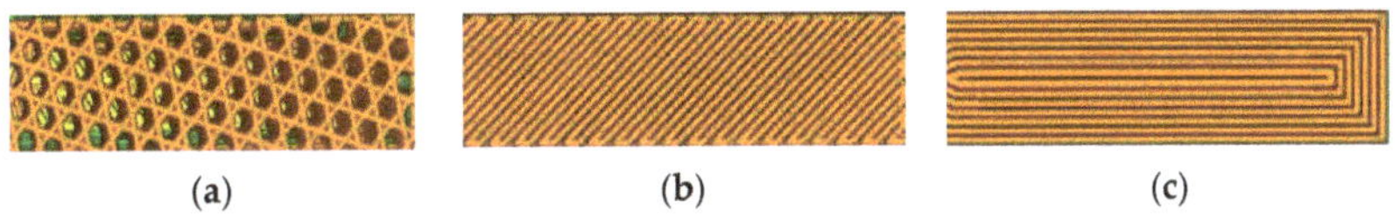

Figure 1. The profiles of internal geometries: Tri-hexagonal (**a**), Zigzag (**b**) and Concentric (**c**).

The experimental procedure involves two main parts: 3D printing and fatigue testing. 3D printing is performed using FDM Ultimaker S3 and S5 printers (Utrecht, Netherlands) whereas fatigue tests are conducted utilizing an SM1090V rotating bending fatigue machine (TecQuipment Ltd. Nottingham, United Kingdom) that can work at frequencies up to 60 Hz. By so doing, the stress ratio was −1. All the applied stress amplitudes were estimated as follows:

$$\sigma_a = \frac{32\,l\,F}{\pi d^3} \quad (1)$$

where l (approximately 28 mm) is the distance from the specimen shoulders where the load was applied to the gauge section, d (about 4 mm) is the gauge section diameter, and F is the applied force on the shoulders of the specimens (Figure 2).

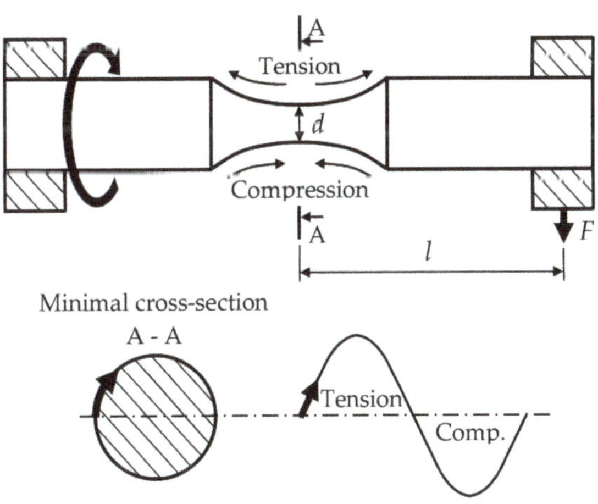

Figure 2. Illustration of stress realization by SM1090V rotating bending fatigue machine.

The observed fatigue results were analyzed quantitatively by fitting the fatigue data in terms of nominal stress amplitude and the number of cycles to failure with a Basquin equation in the form of $\sigma_a = A(N)^b$ where N is the fatigue life. When the curves are plotted on a log-log scale, they can be represented as regression lines. Then, log(A) can be determined as the intercept of the linear fit and its gradient represents b.

After the printing process, the dimensions of a part were measured using a caliper because they can be modified after being printed. The hourglass specimen with 64 mm in length and 9 mm in diameter was used to fit the machine dimensions. The geometry is illustrated in Figure 3.

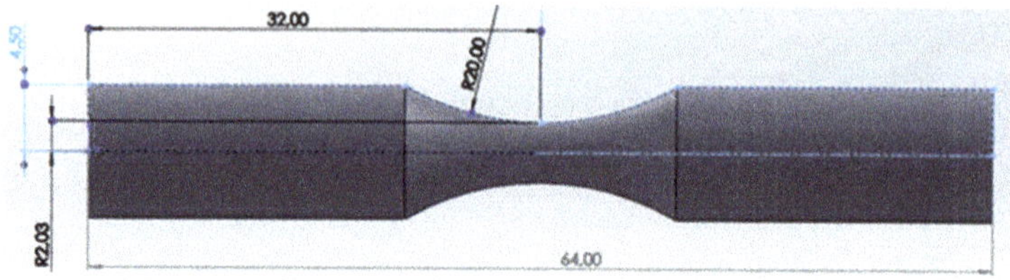

Figure 3. Specimen dimension for fatigue tests.

Test results were analyzed using Taguchi analysis. Such a technique is applied to study the effect of nozzle diameter, internal geometric structure, and printing speed on the fatigue life of the materials studied. ANCOVA was used for a multiple regression analysis in which there is at least one quantitative and one categorical variable. And by doing this, the categorical variable with three kinds of internal geometries was re-coded as two new columns (one of the variables was dropped to avoid multi-collinearity) with 0 and 1. The variables were coded 0 for any case that did not match the variable name and 1 for any case that did match the variable name.

During the numerical part of the study, FE analysis was carried out to simulate the test, and simulation results were obtained and compared with experimental results. A numerical study was conducted in Ansys 2021 R2 Workbench static structural analysis. The specimens were designed as a single solid structure. To replicate tests, the specimen was fixed on one end and the force was applied to another end to create the bending. Mesh parameters include 14,268 elements, 61,479 nodes, and an element size of 0.8 mm. The material properties (S-N curves) used in the analysis were taken from work [37]. The methodology steps that were followed during the study are presented in Figure 4.

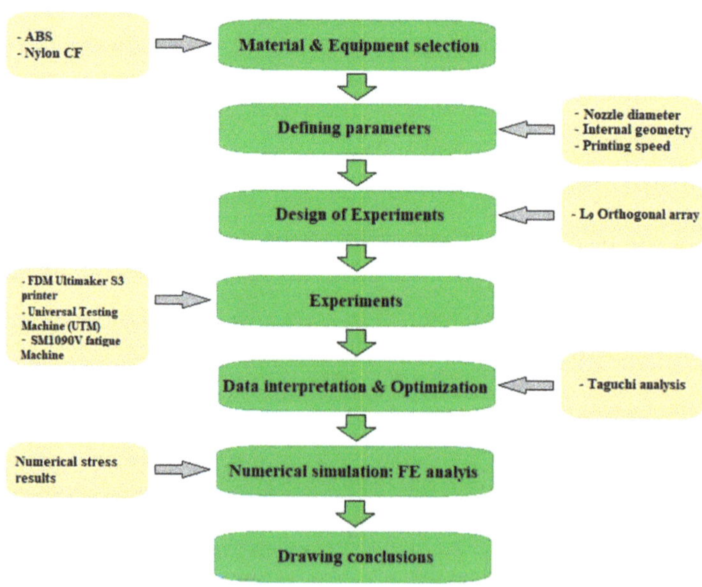

Figure 4. Methodology steps.

3. Results and Discussion

3.1. Experiment

To find the S-N curve, stress was applied by gradually changing the load from 95% to 80% of the ultimate tensile strength (UTS) of materials. In order to obtain UTS, nine tensile tests were conducted for each ABS and Nylon material at different configurations of parameters. The specimen geometry used in tensile tests was designed according to ASTM D638 Type I standard with a working length of 57 mm, work width of 13 mm, and thickness of 3.2 mm. As a result, the mean UTS values were 28.1 ± 2.9 and 50.6 ± 4 MPa for ABS and Nylon, respectively.

Fatigue test results were obtained in terms of the Stress Amplitude vs. Fatigue Life diagrams (Figures 5 and 6). Overall, the experimental results showed that Nylon performed better than ABS: the fatigue strength of Nylon is higher than that of ABS at the same values of the fatigue life. Specimens made of Nylon material could withstand about 45 MPa load under 9000 cycles, while ABS specimens could withstand about 26.5 MPa under 9000 cycles. However, taking into account the UTS of materials, one can notice a significant lead in the fatigue performance of ABS specimens. Specifically, when 95% load of each material's UTS was applied, the ABS part could endure 8200–9400 rotations, whereas the Nylon part at its best configuration could not exceed 6720 revolutions. The same trend can be observed for 90%, 85%, and 80% of UTS loads. In addition, fatigue tests showed which configurations have higher fatigue performance and which parameters have a higher influence on results. On average for all loads, for both Nylon and ABS materials, 111 configurations showed the highest results, while 332 and 321 configurations appeared to be the least resistant to fatigue failure. 3D-printed specimens after fatigue tests are presented in Figure 7.

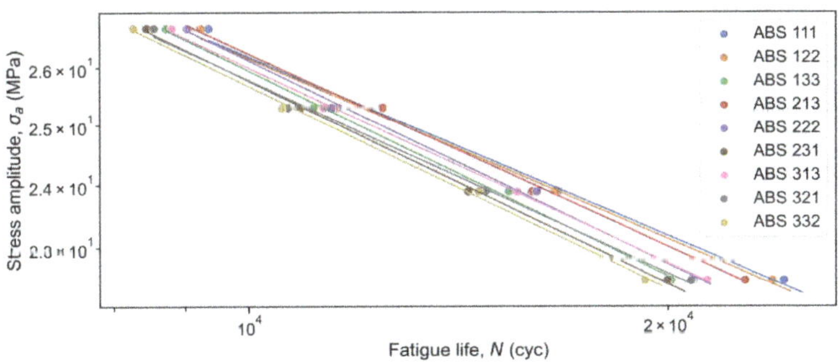

Figure 5. Experimental S-N curve for ABS.

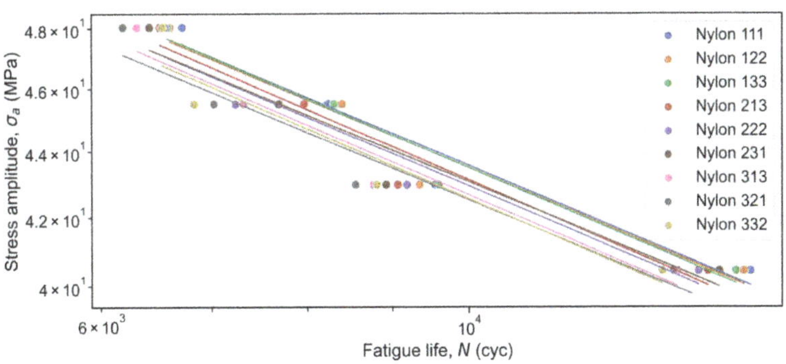

Figure 6. Experimental S-N curve for Nylon.

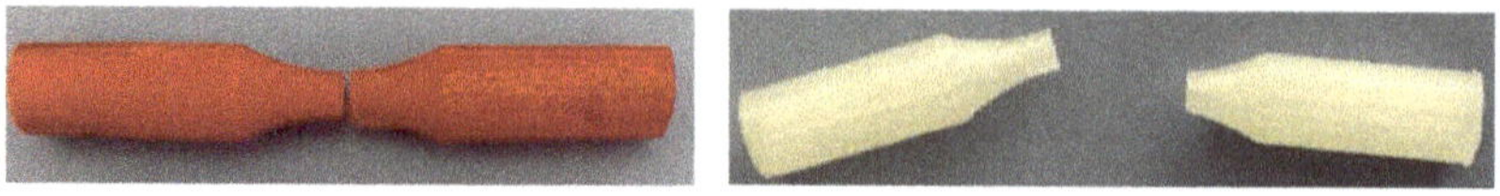

(**a**) (**b**)

Figure 7. Tested 3D-printed fatigue specimens for ABS (**a**) and Nylon (**b**).

The obtained Basquin parameters were extracted from the plots and reported for each manufacturing condition in Table 5. Smaller absolute b values (i.e., reduced slope) represent a higher fatigue performance. As one may see from Table 5, the Nylon material has higher log(A) and b values than ABS overall. It means that initially Nylon has larger fatigue strength but loses it more intensively than ABS.

Table 5. Basquin's parameters.

Test Code	ABS		Nylon	
	log(*A*)	***b***	**log(*A*)**	***b***
111	2.12 ± 0.05	−0.176 ± 0.012	2.50 ± 0.14	−0.215 ± 0.035
122	2.13 ± 0.04	−0.177 ± 0.010	2.50 ± 0.15	−0.214 ± 0.038
133	2.22 ± 0.03	−0.202 ± 0.007	2.51 ± 0.12	−0.218 ± 0.030
213	2.18 ± 0.03	−0.190 ± 0.008	2.53 ± 0.15	−0.223 ± 0.038
222	2.22 ± 0.04	−0.200 ± 0.009	2.50 ± 0.15	−0.217 ± 0.039
231	2.21 ± 0.02	−0.200 ± 0.005	2.48 ± 0.15	−0.211 ± 0.039
313	2.18 ± 0.02	−0.192 ± 0.004	2.51 ± 0.15	−0.221 ± 0.037
321	2.17 ± 0.04	−0.190 ± 0.009	2.48 ± 0.16	−0.214 ± 0.040
332	2.21 ± 0.03	−0.200 ± 0.006	2.52 ± 0.19	−0.222 ± 0.048

3.2. *Sensitivity Analysis*

As discussed in the previous section, the Taguchi method was applied to our experiments to analyze the effect of each printing parameter on the final result without having to execute the full factorial experiment that would have 3 to the power of 3, which is 27 different tests. In Table 6, the mean value of each final result, consisting of the logarithm of the fatigue life for ABS and Nylon, is calculated according to the experiment level of changing parameters. The delta is the difference between the maximum and minimum mean values of the corresponding category. According to this delta value, the ranking was made to conclude this section of the sensitivity analysis. The fatigue tests were carried out at four levels of the stress amplitude. Therefore, for the sensitivity analysis, it was used the average value of the fatigue life as follows:

$$N_m = \frac{1}{n}\log\left(\prod_{i=1}^{n} N_i\right) \tag{2}$$

where n is the number of stress amplitude values (n = 4).

Table 6. The comparison of mean values of data according to their categories and ranking of parameters for ABS and Nylon materials.

Changing Parameter	Nozzle Diameter	Internal Geometry	Printing Speed
Level:		The logarithm of fatigue life | ABS	
1	4.1450	4.1478	4.1260
2	4.1313	4.1343	4.1295
3	4.1131	4.1074	4.1340
Delta	0.0319	0.0403	0.0081
Rank	**2**	**1**	**3**

Table 6. *Cont.*

Changing Parameter	Nozzle Diameter	Internal Geometry	Printing Speed
Level:	The logarithm of fatigue life I Nylon		
1	3.9712	3.9535	3.9486
2	3.9492	3.9460	3.9475
3	3.9281	3.9490	3.9525
Delta	0.0431	0.0075	0.0050
Rank	**1**	**2**	**3**

Sensitivity analysis revealed that internal geometry had the highest influence on fatigue performance of FDM printed parts made of ABS whereas it became the second most important parameter for Nylon. As can be seen from Figure 8, ABS with 'concentric' shape and Nylon with 'zigzag' internal geometries perform in fatigue tests significantly worse by margin than the other two geometries. The 'tri-hexagon' structure resulted in the highest fatigue life. This internal structure enables more dense crisscrossed supporting offsets that help layers stick to each other and form support. In the case of the 'zigzag' structure, it has less dense crisscrossing offset numbers and thus weaker support. Moreover, walls created by 'zigzag' and 'tri-hexagonal' infill patterns enable the transfer of a certain portion of the stress to neighboring walls, thus decreasing stress concentrations at local points and delaying crack propagation [33].

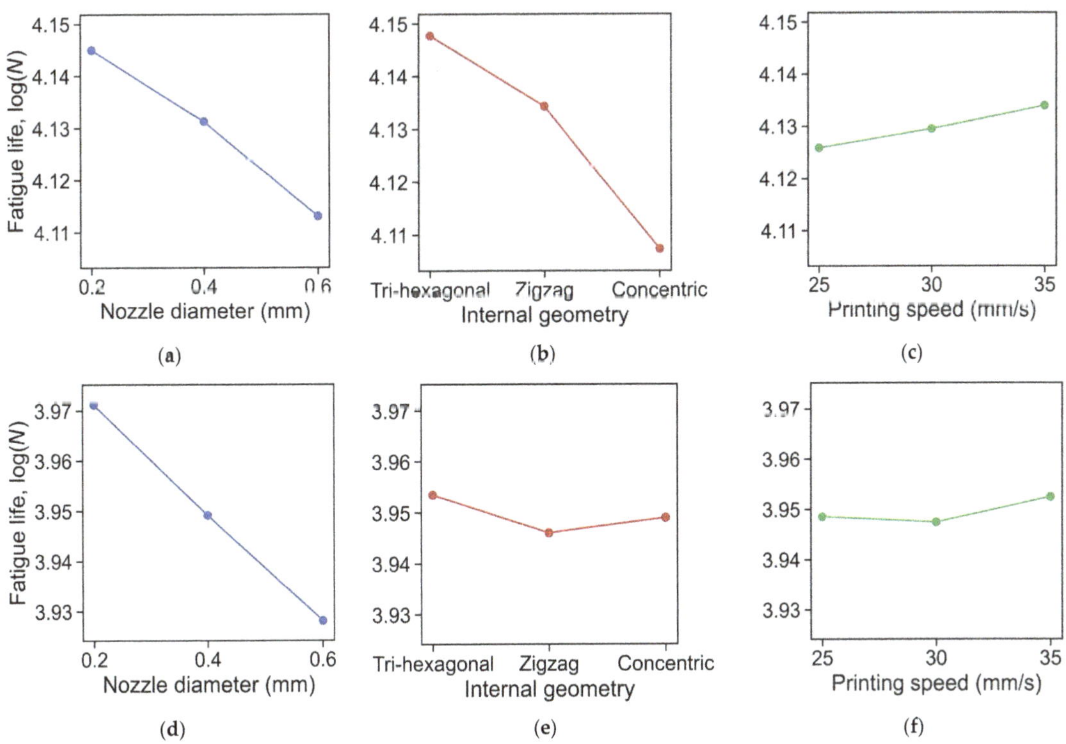

Figure 8. Sensitivity analysis of ABS (**a**–**c**) and Nylon (**d**–**f**) fatigue test for Nozzle diameter (**a**,**d**), Internal geometry (**b**,**e**), and Printing speed (**c**,**f**).

Additionally, obtained results proved that a change in nozzle diameter had a notable influence on the continuity of the printed specimen and hence on mechanical properties.

An increase in nozzle diameter leads to a reduction of continuity of the internal volume of the part and consecutively to a decrease in fatigue life. Using nozzles with small diameters ensures high density and packing of more material in a certain volume and thus notably reduces the volume of air-filled cavities.

Compared with other printing parameters, the effect of change in printing speed on fatigue performance appears to be much lower. On top of that, from the obtained sensitivity analysis we cannot distinguish a clear pattern or trends associated with printing speed. For different loads and different materials, an increase in printing speed leads to different results without yielding any tendencies. One possible reason for such results might be the small ranges selected for printing speeds [38]. It varied just from 25 to 35 mm/s. Perhaps, the fatigue properties will be more sensitive if a larger speed range were to be chosen, which needs further investigation.

3.3. ANCOVA Regression Analysis

In this research study, ANCOVA multiple regression analysis was applied [39]. The model used is the additive one, i.e., it does not take into account any interaction effects. From Table 7 one can see that the model explains 99.2% and 93.2% of the variability in test scores for ABS and for Nylon, respectively (adjusted coefficient of determination, R^2, is 0.992 and 0.932), and the standard error of estimate represents how far data fall from the regression predictions (fatigue life values in terms of logarithmic scale). Moreover, the F statistic 'F-value' of 842.4 and 97.4 with *p*-values of less than 0.001 shows the advantage of this model over an intercept-only model that predicts the average output for all the data.

Table 7. Model summary results for fatigue life in terms of logarithmic scale.

Material	R^2	Adjusted R^2	Std. Error of Estimate	F-Value	*p*-Value
ABS	0.993	0.992	0.0135	842.4	<0.001 ***
Nylon	0.942	0.932	0.0331	97.40	<0.001 ***

Significance levels: *** *p*-val. $\leq$ 0.001 (significant), ** *p*-val. $\leq$ 0.01 (very significant), * *p*-val. $\leq$ 0.05 (highly significant).

Reviewing the regression results in Tables 7 and 8, the models can be represented as:

$$\textbf{ABS: } \log(N) = 11.39 - 0.028 \times \text{Con.geom.} + 0.012 \times \text{Tri.geom.} - 5.20 \times \log(\sigma_a) - 0.08 \times \text{Noz.d.} + 0.0004 \times \text{Pr.sp.} \quad (3)$$

$$\textbf{Nylon: } \log(N) = 11.12 + 0.004 \times \text{Tri.geom.} - 0.003 \times \text{Zig.geom.} - 4.34 \times \log(\sigma_a) - 0.108 \times \text{Noz.d.} + 0.0002 \times \text{Pr.sp.} \quad (4)$$

The simplified model equations are shown here as follows:

$$\textbf{ABS \& Zigzag: } \log(N) = 11.39 - 5.20 \times \log(\sigma_a) - 0.08 \times \text{Noz.d.} + 0.0004 \times \text{Pr.sp.} \quad (5)$$

$$\textbf{ABS \& Concentric: } \log(N) = 11.39 - 0.028 - 5.20 \times \log(\sigma_a) - 0.08 \times \text{Noz.d.} + 0.0004 \times \text{Pr.sp.} \quad (6)$$

$$\textbf{ABS \& Tri-hexagonal: } \log(N) = 11.39 + 0.012 - 5.20 \times \log(\sigma_a) - 0.08 \times \text{Noz.d.} + 0.0004 \times \text{Pr.sp.} \quad (7)$$

$$\textbf{Nylon \& Concentric: } \log(N) = 11.12 - 4.34 \times \log(\sigma_a) - 0.108 \times \text{Noz.d.} + 0.0002 \times \text{Pr.sp.} \quad (8)$$

$$\textbf{Nylon \& Zigzag: } \log(N) = 11.12 - 0.003 - 4.34 \times \log(\sigma_a) - 0.108 \times \text{Noz.d.} + 0.0002 \times \text{Pr.sp.} \quad (9)$$

$$\textbf{Nylon \& Tri-hexagonal: } \log(N) = 11.12 + 0.004 - 4.34 \times \log(\sigma_a) - 0.108 \times \text{Noz.d.} + 0.0002 \times \text{Pr.sp.} \quad (10)$$

Table 8. Multiple linear regression results for fatigue life in terms of logarithmic scale.

Materials	Coefficient	Std. Err.	t-Value	p-Value	Confidence Interval	
					[0.025	0.975]
ABS						
Const.	11.39	0.11	99.8	<0.001 ***	11.15	11.62
$\log(\sigma_a)$	−5.20	0.08	−64.2	<0.001 ***	−5.37	−5.04
Conc.geom.	−0.028	0.006	−4.9	<0.001 ***	−0.039	−0.016
Tri-hex.geom.	0.012	0.006	2.1	0.049 *	6.6×10^{-5}	0.024
Noz.diam.	−0.080	0.014	−5.8	<0.001 ***	−0.108	−0.052
Print.speed	0.0004	0.001	0.7	0.495	−0.001	0.002
Nylon						
Const.	11.12	0.33	33.7	<0.001 ***	10.45	11.79
$\log(\sigma_a)$	−4.34	0.20	−21.8	<0.001 ***	−4.74	−3.93
Tri-hex.geom.	0.004	0.014	0.3	0.761	−0.024	0.032
Zigz.geom.	−0.003	0.014	−0.2	0.845	−0.031	0.025
Noz.diam.	−0.108	0.034	−3.2	0.003 **	−0.177	−0.039
Print.speed	0.0002	0.001	0.1	0.912	−0.003	0.003

Significance levels: *** p-val. $\leq$ 0.001 (significant), ** p-val. $\leq$ 0.01 (very significant), * p-val. $\leq$ 0.05 (highly significant).

From the equations, one can see that the slope of $\log(N)$ vs. $\log(\sigma_a)$ is −5.20 and −4.34 for ABS and Nylon. Meanwhile, the intercepts of 11.39 (with a 95% confidence interval from 11.15 to 11.62) and 11.12 (with a 95% confidence interval from 10.45 to 11.79) are the expected mean value of $\log(N)$ of the ABS with the 'Zigzag' internal geometry and Nylon with the 'Concentric' geometry. The intercept depends on the internal geometry: the 'Tri-hexagonal' geometry increases the intercept by 0.012 for ABS and by 0.004 for Nylon whereas the 'Concentric' and 'Zigzag' geometries decrease it by 0.028 and 0.003 for ABS and Nylon materials, respectively. However, the geometry effect is statistically significant only for ABS material (conc.geom.: t-value = −4.9, p-value < 0.001; tri-hex.geom.: t-value = 2.1, p-value = 0.049).

Meanwhile, there is a statistically significant effect of the nozzle diameter on fatigue life: the slope $\log(N)$ vs. Noz.diam. of −0.080 with t-value = −5.8 and p-value < 0.001 for ABS as well as −0.108 with t-value = −3.2 and p-value = 0.003 for Nylon. It means the mean value of $\log(N)$ decreases by 0.008 and 0.0108 for every 0.1 mm magnification of the nozzle diameter for ABS and Nylon, respectively. On the contrary, the printing speed in the studied range from 24 to 35 mm/s negligibly affects the result for both materials. It should be noted that all the obtained results are in good agreement with the previous section.

In addition, it should be noted that at least 15 tests with 13 degrees of freedom, respectively are necessary according to fatigue standards (for example, PN-H-04325:1976) [40]. In this study, if one considers every S-N curve separately, it looks like four points are not enough to plot a reliable line of best fit because there are only 2 degrees of freedom. However, an application of ANCOVA regression analysis allows for considering test points not separately for every S-N curve but as a single dataset with 36 experiments for an additive regression model (Equation (3) or (4)) with six constants (Table 8), and 30 degrees of freedom. According to the ASTM E-739-91, it is enough for preliminary and exploratory tests [40]. By doing so, one also can calculate the coefficient of determination and the standard error (how far data fall from the predictions, Table 7) that shows the fatigue life variability in terms of logarithmic scale. Thus, the authors assume that 30 tests with 36 degrees of freedom are enough to confirm the trends shown in this study from a statistical viewpoint. In order to use the optimal printing parameters to calculate the fatigue behavior of real structures in the future, the S-N curve constants have to be computed more accurately.

3.4. Numerical Simulation

A numerical study for ABS and PA6 (Nylon) plastics is presented in Figures 9 and 10, and Table 9. Comparing the parameters of configured parts with simulation results revealed that the fatigue life of the printed part was on average 11.4% higher than of the numerical one for ABS plastic in terms of logarithmic scale. According to static structural analysis, the ABS part can withstand $10^{3.64}$ cycles under 90% UTS (25.3 MPa), while experimental results show $10^{4.08}$ revolutions at the same load level (Figure 9, Table 9). For Nylon material at 90% UTS cyclic load (45.5 MPa), numerical fatigue life reached $10^{3.75}$ rotations, whereas the experimental one could reach $10^{3.92}$ cycles (Figure 10, Table 9). The difference might take place due to a number of simplifications and limitations undertaken during numerical analysis. Firstly, simulations consider the specimen as a solid structure. In practice, internal structure plays a significant role in the mechanical properties of the specimen. More realistic models (for instance, layered or laminate models) could provide more accurate simulated results (with differences lower than the reported ones). Moreover, voids, cracks, and internal defects may also notably decrease the performance of the parts during tests. Despite up to 14% differences between numerical and experimental results, for the most part, static structural fatigue analysis showed valid results.

Table 9. Comparison of experimental results and numerically obtained data.

UTS %	ABS			Nylon		
	log(*N*)		Difference	log(*N*)		Difference
	Experiment	Numerical	%	Experiment	Numerical	%
95	3.97	3.56	10.5	3.83	3.30	13.8
90	4.08	3.64	10.9	3.92	3.75	4.2
85	4.22	3.74	11.5	3.98	4.26	−7.1
80	4.39	3.84	12.6	4.17	4.73	−13.4

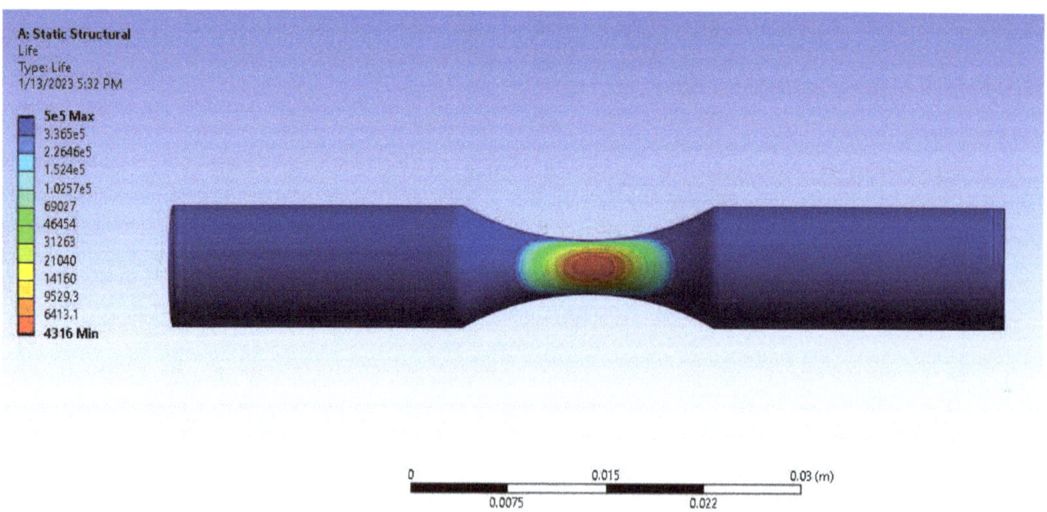

Figure 9. ABS numerical fatigue analysis of fatigue life under 90% UTS load.

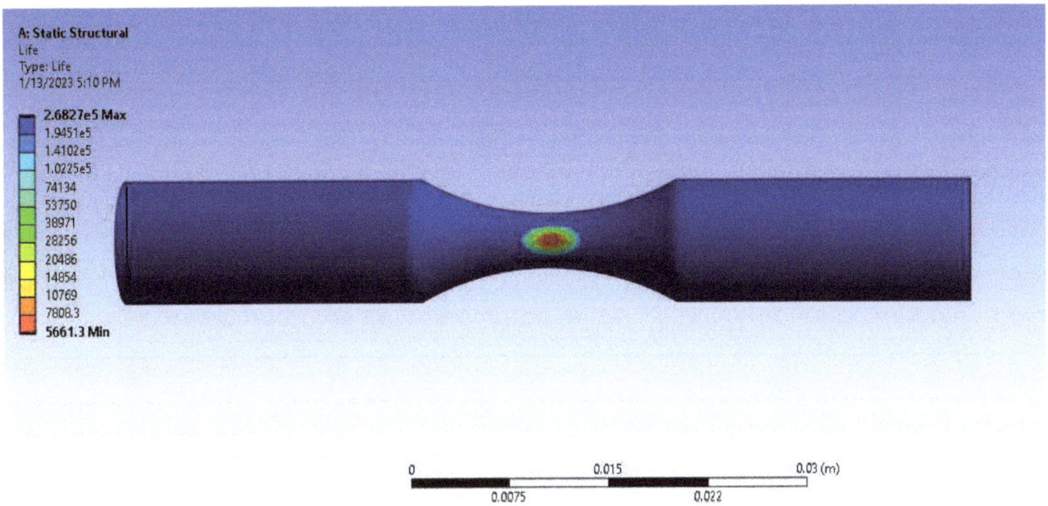

Figure 10. Nylon numerical fatigue analysis of fatigue life under 90% UTS load.

4. Conclusions

The parametric study of the printing parameters on the fatigue properties of FDM-printed ABS and Nyon applying Taguchi analysis was developed. In this regard, the work focused on the change in fatigue life due to the change in the printing parameters in a chosen range. The change range of the parameters was based on the conclusion of the review of the literature.

A comparative analysis of the experiments showed that under repetitive tensile and compression load, Nylon performed better than ABS. However, ABS has a lower slope, i.e., fatigue strength decreases less intensively with increasing fatigue life than that for Nylon.

Sensitivity analysis indicated that internal geometry had the highest influence on the fatigue performance of FDM printed parts made of ABS and it became the second most important parameter for Nylon. By so doing, the 'tri-hexagon' structure and nozzle diameter of 0.2 mm resulted in the highest fatigue life for both materials. On the contrary, the printing speed parameter in the range of 25–35 mm/s had the least effect on the fatigue life values. The comparison of the experimental and numerical values with theoretical ones showed small differences.

The ANCOVA analysis for multiple linear regression confirmed that 'Tri-hexagonal' geometry had a positive effect on fatigue life. However, the effect was statistically significant only for ABS. For Nylon, the change in geometry might be considered negligible. Meanwhile, there is a statistically significant effect of the nozzle diameter on the fatigue life: the mean value of $\log(N)$ decreases by 0.008 and 0.0108 for every 0.1 mm magnification of the nozzle diameter for ABS and Nylon, respectively. Conversely, the printing speed change in a narrow range (25–35 mm/s) negligibly affects the result for both materials. However, it should be noted that the fatigue properties will probably be more sensitive to a larger speed range.

Thus, the application of results could be served as the optimization factors for better mechanical properties of FDM printed parts for three different types of tests. The findings can serve as reference data for future parametric studies and in the search for a suitable printing configuration for different plastic applications.

Author Contributions: Conceptualization, A.P., D.T.; methodology, Y.A., S.D., G.S., A.T. and A.Y.; software, Y.A., A.T. and A.Y.; validation, S.D., G.S. and A.Y.; formal analysis, Y.A., S.D, G.S., A.T. and A.Y.; investigation, Y.A., S.D., G.S., A.T. and A.Y.; resources, A.P. and D.T.; data curation, Y.A., S.D., G.S., A.T. and A.Y.; writing—original draft preparation, A.Y. and A.P.; writing—review and editing, A.P. and D.T.; visualization, A.P. and D.T.; supervision, A.P. and D.T.; project administration, D.T. and A.P.; funding acquisition, D.T. All authors have read and agreed to the published version of the manuscript.

Funding: This research was funded under the target program (no. OR07665556) for project entitled "Additive production systems and metal powders for Kazakhstan industry", by the Ministry of Industry and Infrastructure Development of the Republic of Kazakhstan.

Data Availability Statement: The data will be provided by the corresponding author upon request.

Conflicts of Interest: The authors declare no conflict of interest.

Abbreviations

Nomenclature	Description
l	the distance from the specimen shoulders where the load was applied to the gauge section
d	the gauge section diameter
F	the applied force on the shoulders of the specimens
N	the fatigue life
$\log(A)$	the intercept of the linear fit
b	the gradient of the linear fit
σ_a	the stress amplitude
UTS	the ultimate tensile strength
FDM	the Fused Deposition Modeling
ABS	acrylonitrile butadiene styrene
Nylon	Polyamide

References

1. Mercado-Colmenero, J.M.; Martin-Doñate, C.; Moramarco, V.; Attolico, M.A.; Renna, G.; Rodriguez-Santiago, M.; Casavola, C. Mechanical Characterization of the Plastic Material GF-PA6 Manufactured Using FDM Technology for a Compression Uniaxial Stress Field via an Experimental and Numerical Analysis. *Polymers* **2020**, *12*, 246. [CrossRef]
2. Swetham, T.; Reddy, K.M.; Huggi, A.; Kumar, M.N. A Critical Review on of 3D Printing Materials and Details of Materials used in FDM. *Int. J. Sci. Res. Sci. Eng. Technol.* **2017**, *3*, 353–361.
3. Salem Bala, A.; bin Wahab, S.; binti Ahmad, M. Elements and Materials Improve the FDM Products: A Review. *Adv. Eng. Forum* **2016**, *16*, 33–51. [CrossRef]
4. Popescu, D.; Zapciu, A.; Amza, C.; Baciu, F.; Marinescu, R. FDM process parameters influence over the mechanical properties of polymer specimens: A review. *Polym. Test.* **2018**, *69*, 157–166. [CrossRef]
5. Dey, A.; Yodo, N. A Systematic Survey of FDM Process Parameter Optimization and Their Influence on Part Characteristics. *J. Manuf. Mater. Process* **2019**, *3*, 64. [CrossRef]
6. Formlabs. Guide to 3D Printing Materials: Types, Applications, and Properties. Available online: https://formlabs.com/blog/3d-printing-materials/ (accessed on 14 November 2021).
7. Rouf, S.; Malik, A.; Singh, N.; Raina, A.; Naveed, N.; Siddiqui, M.I.H.; Haq, M.I.U. Additive manufacturing technologies: Industrial and medical applications. *Sustain. Oper. Comput.* **2022**, *3*, 258–274. [CrossRef]
8. Li, L.; Sun, Q.; Bellehumeur, C.; Gu, P. Composite modeling and analysis for fabrication of FDM prototypes with locally controlled properties. *J. Manuf. Process.* **2002**, *4*, 129–141. [CrossRef]
9. Tao, Y.; Kong, F.; Li, Z.; Zhang, J.; Zhao, X.; Yin, Q.; Xing, D.; Li, P. A review on voids of 3D printed parts by fused filament fabrication. *J. Mater. Res. Technol.* **2021**, *15*, 4860–4879. [CrossRef]
10. Solomon, I.J.; Sevvel, P.; Gunasekaran, J. A review on the various processing parameters in FDM. *Mater. Today Proc.* **2020**, *37*, 509–514. [CrossRef]
11. Syrlybayev, D.; Zharylkassyn, B.; Seisekulova, A.; Akhmetov, M.; Perveen, A.; Talamona, D. Optimisation of Strength Properties of FDM Printed Parts—A Critical Review. *Polymers* **2021**, *13*, 1587. [CrossRef]
12. Vicente, C.M.S.; Martins, T.S.; Leite, M.; Ribeiro, A.; Reis, L. Influence of fused deposition modeling parameters on the mechanical properties of ABS parts. *Polym. Adv. Technol.* **2020**, *31*, 501–507. [CrossRef]
13. Ramesh, M.; Panneerselvam, K. Mechanical investigation and optimization of parameter selection for Nylon material processed by FDM. *Mater. Today Proc.* **2020**, *46*, 9303–9307. [CrossRef]

14. Alafaghani, A.; Qattawi, A.; Alrawi, B.; Guzman, A. Experimental Optimization of Fused Deposition Modelling Processing Parameters: A Design-for-Manufacturing Approach. *Procedia Manuf.* **2017**, *10*, 791–803. [CrossRef]
15. de Toro, E.V.; Sobrino, J.C.; Martínez, A.M.; Eguía, V.M. Analysis of the influence of the variables of the Fused Deposition Modeling (FDM) process on the mechanical properties of a carbon fiber-reinforced polyamide. *Procedia Manuf.* **2019**, *41*, 731–738. [CrossRef]
16. Garg, A.; Bhattacharya, A. An insight to the failure of FDM parts under tensile loading: Finite Element Analysis and experimental study. *Int. J. Mech. Sci.* **2016**, *120*, 225–236. [CrossRef]
17. Mercado-Colmenero, J.M.; La Rubia, M.D.; Mata-Garcia, E.; Rodriguez-Santiago, M.; Martin-Doñate, C. Experimental and Numerical Analysis for the Mechanical Characterization of PETG Polymers Manufactured with FDM Technology under Pure Uniaxial Compression Stress States for Architectural Applications. *Polymers* **2020**, *12*, 2202. [CrossRef]
18. Mohamed, O.A.; Masood, S.H.; Bhowmik, J.L. Optimization of fused deposition modeling process parameters: A review of current research and future prospects. *Adv. Manuf.* **2015**, *3*, 42–53. [CrossRef]
19. Sheoran, A.J.; Kumar, H. Fused Deposition modeling process parameters optimization and effect on mechanical properties and part quality: Review and reflection on present research. *Mater. Today Proc.* **2020**, *21*, 1659–1672. [CrossRef]
20. Jerez-Mesa, R.; Travieso-Rodriguez, J.A.; Llumà-Fuentes, J.; Gomez-Gras, G.; Puig, D. Fatigue lifespan study of PLA parts obtained by additive manufacturing. *Procedia Manuf.* **2017**, *13*, 872–879. [CrossRef]
21. Gomez-Gras, G.; Jerez-Mesa, R.; Travieso-Rodriguez, J.A.; Lluma-Fuentes, J. Fatigue performance of fused filament fabrication PLA specimens. *Mater. Des.* **2018**, *140*, 278–285. [CrossRef]
22. Pertuz, A.D.; Díaz-Cardona, S.; González-Estrada, O.A. Static and fatigue behaviour of continuous fibre reinforced thermoplastic composites manufactured by fused deposition modelling technique. *Int. J. Fatigue* **2020**, *130*, 105275. [CrossRef]
23. Azadi, M.; Dadashi, A.; Dezianian, S.; Kianifar, M.; Torkaman, S.; Chiyani, M. High-cycle bending fatigue properties of additive-manufactured ABS and PLA polymers fabricated by fused deposition modeling 3D-printing. *Forces Mech.* **2021**, *3*, 100016. [CrossRef]
24. Afrose, M.F.; Masood, S.H.; Iovenitti, P.; Nikzad, M.; Sbarski, I. Effects of part build orientations on fatigue behaviour of FDM-processed PLA material. *Prog Addit Manuf* **2016**, *1*, 21–28. [CrossRef]
25. Lee, J.; Huang, A. Fatigue analysis of FDM materials. *Rapid Prototyp. J.* **2013**, *19*, 291–299. [CrossRef]
26. Terekhina, S.; Tarasova, T.; Egorov, S.; Skornyakov, I.; Guillaumat, L.; Hattali, M.L. The effect of build orientation on both flexural quasi-static and fatigue behaviours of filament deposited PA6 polymer. *Int. J. Fatigue* **2020**, *140*, 105825. [CrossRef]
27. Letcher, T.; Waytashek, M. Material Property Testing of 3D-Printed Specimen in PLA on an Entry-Level 3D Printer. In Proceedings of the ASME 2014 International Mechanical Engineering Congress and Exposition, Montreal, QC, Canada, 14–20 November 2014; Volume 2A, pp. 14–20. [CrossRef]
28. Ziemian, S.; Okwara, M.; Ziemian, C.W. Tensile and fatigue behavior of layered acrylonitrile butadiene styrene. *Rapid Prototyp. J.* **2015**, *21*, 270–278. [CrossRef]
29. Ziemian, C.W.; Ziemian, R.D.; Haile, K.V. Characterization of stiffness degradation caused by fatigue damage of additive manufactured parts. *Mater. Des.* **2016**, *109*, 209–218. [CrossRef]
30. Terekhina, S.; Tarasova, T.; Egorov, S.; Guillaumat, L.; Hattali, M.L. On the difference in material structure and fatigue properties of polyamide specimens produced by fused filament fabrication and selective laser sintering. *Int. J. Adv. Manuf. Technol.* **2020**, *111*, 93–107. [CrossRef]
31. Svetlana, T.; Tatiana, T.; Sergei, E.; Innokentiy, S.; Laurent, G.; Lamine, H. Flexural quasi-static and fatigue behaviours of fused filament deposited PA6 and PA12 polymers. *Int. J. Adv. Manuf. Technol.* **2021**, *117*, 2041–2048. [CrossRef]
32. Miller, A.T.; Safranski, D.L.; Smith, K.E.; Sycks, D.G.; Guldberg, R.E.; Gall, K. Fatigue of injection molded and 3D printed polycarbonate urethane in solution. *Polymer* **2017**, *108*, 121–134. [CrossRef]
33. Ferreira, C.M.; Vicente, C.M.S.; Sardinha, M.; Leite, M.; Reis, L. Characterization of 3D printed ABS specimens under static and cyclic torsional loadings. *Procedia Struct. Integr.* **2021**, *34*, 205–210. [CrossRef]
34. Zhang, H. Characterization of Tensile, Creep, and Fatigue Properties of 3D Printed Acrylonitrile Butadiene Styrene. Master's Thesis, Purdue University, West Lafayette, IN, USA, 2016. [CrossRef]
35. Safai, L.; Cuellar, J.S.; Smit, G.; Zadpoor, A.A. A review of the fatigue behavior of 3D printed polymers. *Addit. Manuf.* **2019**, *28*, 87–97. [CrossRef]
36. Radhwan, H.; Shayfull, Z. Optimization Parameter Effects on the Strength of 3D-Printing Process using Taguchi method. *AIP Conf. Proc.* **2019**, *2129*, 020154. [CrossRef]
37. Lu, Z.; Feng, B.; Loh, C. Fatigue behaviour and mean stress effect of thermoplastic polymers and composites. *Frat. Ed Integrità Strutt.* **2018**, *46*, 150–157. [CrossRef]
38. Verbeeten, W.M.; Arnold-Bik, R.J.; Lorenzo-Bañuelos, M. Print velocity effects on strain-rate sensitivity of acrylonitrile-butadiene-styrene using material extrusion additive manufacturing. *Polymers* **2021**, *13*, 149. [CrossRef] [PubMed]

39. Seltman, H.J. Experimental Design and Analysis. Carnegie Mellon University: Pittsburgh, PA, USA, 2018; 428p.
40. Strzelecki, P.; Sempruch, J. Experimental Method for Plotting S-N Curve with a Small Number of Specimens. *Pol. Marit. Res.* **2017**, *23*, 129–137. [CrossRef]

micromachines

MDPI

Review

Research Progress on Bonding Wire for Microelectronic Packaging

Hongliang Zhou *, Yingchong Zhang, Jun Cao, Chenghao Su, Chong Li, Andong Chang and Bin An

School of Mechanical and Power Engineering, Henan Polytechnic University, Jiaozuo 454000, China
* Correspondence: zhlhpu@hpu.edu.cn

Abstract: Wire bonding is still the most popular chip interconnect technology in microelectronic packaging and will not be replaced by other interconnect methods for a long time in the future. Au bonding wire has been a mainstream semiconductor packaging material for many decades due to its unique chemical stability, reliable manufacturing, and operation properties. However, the drastic increasing price of Au bonding wire has motivated the industry to search for alternate bonding materials for use in microelectronic packaging such as Cu and Ag bonding wires. The main benefits of using Cu bonding wire over Au bonding wire are lower material cost, higher electrical and thermal conductivity that enables smaller diameter Cu bonding wire to carry identical current as an Au bonding wire without overheating, and lower reaction rates between Cu and Al that serve to improve the reliability performance in long periods of high temperature storage conditions. However, the high hardness, easy oxidation, and complex bonding process of Cu bonding wire make it not the best alternative for Au bonding wire. Therefore, Ag bonding wire as a new alternative with potential application comes to the packaging market; it has higher thermal conductivity and lower electric resistivity in comparison with Cu bonding wire, which makes it a good candidate for power electronics, and higher elastic modulus and hardness than Au bonding wire, but lower than Cu bonding wire, which makes it easier to bond. This paper begins with a brief introduction about the developing history of bonding wires. Next, manufacturability and reliability of Au, Cu, and Ag bonding wires are introduced. Furthermore, general comparisons on basic performance and applications between the three types of bonding wires are discussed. In the end, developing trends of bonding wire are provided. Hopefully, this review can be regarded as a useful complement to other reviews on wire bonding technology and applications.

Keywords: bonding wire; manufacturability; reliability; general comparison; development trends

Citation: Zhou, H.; Zhang, Y.; Cao, J.; Su, C.; Li, C.; Chang, A.; An, B. Research Progress on Bonding Wire for Microelectronic Packaging. *Micromachines* **2023**, *14*, 432. https://doi.org/10.3390/mi14020432

Academic Editor: Nikolaos Tapoglou

Received: 8 January 2023
Revised: 6 February 2023
Accepted: 9 February 2023
Published: 11 February 2023

1. Introduction

Microelectronics packaging is an essential part of the microelectronics industry, which is one of the pillar industries in countries all over the world. The common chip interconnect technologies in microelectronic packaging include wire bonding [1,2], flip-chip bonding [3,4], tape automated bonding (TAB) [5,6], etc. Wire bonding has been the most cost-effective, mature, and flexible interconnect technology in microelectronic packaging since its invention in the 1960s, which is still used to assemble more than 80% semiconductor packages in the microelectronic packaging industry [7,8]. The bonding wire, an important structural material for microelectronic packaging, plays a vital role in connecting the integrated circuit (IC) chip and the outer lead frame [9]. Statistics show that around 1/4–1/3 of the package failures are caused by wire bonding [10].

Throughout the development history of bonding wires, Au bonding wire as the earliest applied bonding wire has high mechanical strength, excellent oxidation resistance, and simple bonding process. However, owing to the steep increase in Au price and its limited performance development in recent years, the market share of Au bonding wire is decreasing yearly [11,12]. Cu bonding wire, as an alternative wire, has a potential to provide good electrical connections for high-power and highly-integrated electronics because of its higher electrical and thermal conductivity and lower cost than Au [13–15]. However, since the Cu bonding wire has the properties of high hardness and oxidation rate and complex bonding process that will lead to damage of pad during bond, scholars agree that Cu bonding wire is not the ultimate candidate for Au bonding wire [16–18]. Ag bonding wire attracts the attention of scholars because of its excellent electrical and thermal conductivity and lower price than Au. Also, Ag is one of the materials with the most potential to adapt to the development trend of microelectronic packaging with high integration, high density, and high speed, that is considered a promising bonding material for application in the packaging market [19,20]. But there still exists some reliability issues of Ag bonding wire to be solved, so it has not obtained widespread application yet [21]. This paper introduces processing techniques and reliability of the above-mentioned bonding wires, then analyzes and compares basic performance and application between the three bonding wires. Finally, we draw some conclusions and give an outlook on future challenges in microelectronic packaging, which provides technical reference for application and popularization of wire bonding in the packing field.

2. Au Bonding Wire

In the semiconductor manufacturing industry, Au bonding wire is the longest-serving and most widely applied bonding wire with good chemical stability, high processability, high reliability, excellent ductility, electrical conductivity, and low bonding requirements, which plays an important role in power and signal transmission. The quality of Au wire bonding is highly significant to ensure the normal operation of electronic systems [22].

2.1. Au Wire

The purity and diameter of Au wire have influences on its inherent performance. The higher the purity, the better the electrical conductivity and bonding ability, but the lower the elastic modulus and tensile strength, which is unfavorable for the wire loop and strength during the bonding process and will lead to bonding problems such as collapse and tailing, as shown in Figure 1 [23]. The Au wire with a varied portfolio of 2N (99%), 3N (99.9%), 4N (99.99%), etc., has a usual diameter range of 15–50 μm; the smaller the wire diameter, the more difficult the loop control. It is better to choose the purity and diameter of the Au wire according to its working performance requirements in order to ensure the bonding reliability [19]. Papadopoulos et al. [24] carried out a comparative study of 2N and 4N Au wire bond materials based on Automotive Electronics Council Qualification (AEC-Q100). The results showed that the 4N wire is reaching the lift-off values of AEC-Q100 specification at any case and is well suited up to a temperature of 150 °C, thus meeting the automotive requirements. However, the 2N wire seems to have a better performance than 4N wire due to reduced intermetallic phase (IMP) formation growth at higher temperatures.

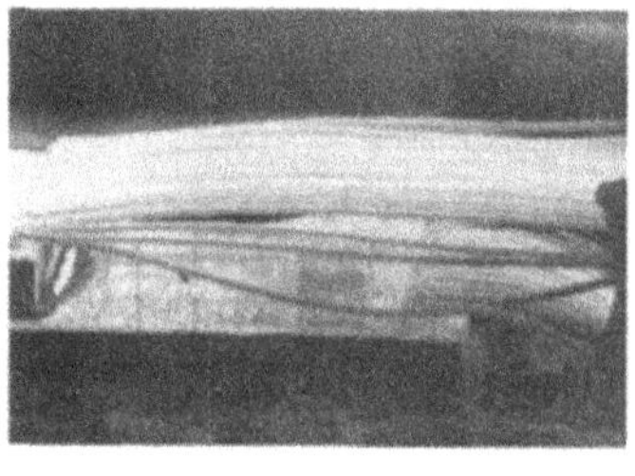

(a)

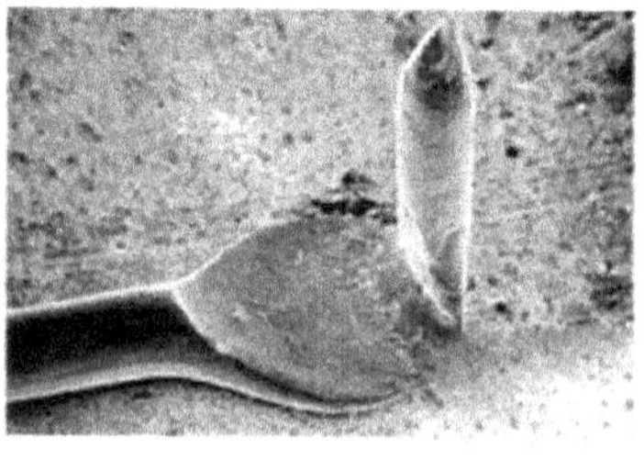

(b)

Figure 1. The phenomenon of wire (**a**) collapse and (**b**) tailing of Au bonding wire [23]. Reproduced with permission from Ref. [23]; published by Precious Metals, 2017.

As IC packaging develops toward miniaturization, high-density, and large-span ratio by leaps and bounds, higher demands are placed on the wire diameter, hardness, and high-temperature performance of Au wire. It is found that the problems of poor heat resistance, low recrystallization temperature, easy generation of harmful intermetallic compounds (IMCs) at high temperature and formation of Kirkendall voids during bonding process, and limited performance development, etc., of Au wire have prevented it fitting the miniaturization requirements of electronic products. Nevertheless, the above problems can be solved by doping trace elements in varied proportions into Au to make Au alloy wire [25,26].

2.2. Au Alloy Wire

In order to keep pace with the development of IC packaging in the direction of miniaturization, adding trace elements is an effective way to increase the strength of Au wire and reduce its wire diameter [25]. Adding Ca, Y, Be, etc., to high-purity Au can increase its recrystallization temperature, and improve its wire strength and loop stability. Adding rare earth elements can make the grain refinement, increase the size of heat-affected zone (HAZ), and improve its high temperature stability; adding Cu, Pd, Pt, etc., can inhibit the growth of IMCs and improve the bonding reliability [26]. Humpston et al. [27] developed a fine wire of the composition Au-1wt%Ti with a diameter of 25 μm through alloying, and its strength is three times higher than that of traditional Au wire. Ag and Au are infinitely miscible in solids and liquids states with a small gap between solidus and liquidus curves [28]; while a great quantity of Ag added to Au will reduce the bonding reliability between the bonding wire and the electrode of the semiconductor device, then alloying elements such as Pd, Rh, and Ru can be doped to improve bonding reliability [29]. In addition, Au alloy wire made with Au and Ni, alkaline earth metals, or rare earth metals has higher strength than traditional Au wire under the same loading conditions [30,31]. Especially, when alkaline earth metals and rare earth metals are added to Au, the strength loss of Au alloy wire is significantly reduced at high temperature [31]. Moreover, studies on the properties and applications of other Au alloy wires such as Au-Ti system, Au-Cu system, Au-Pd system, and Au-Cu-Ca system are reviewed; their advantages compared with traditional Au wire are shown in Table 1 [32,33]. Though the addition of trace elements to Au can improve bonding property of Au alloy wire, its high price due to over 50% mass fraction of Au still restricts its wide use in the global microelectronics packaging market [34,35].

Table 1. Advantages of Au alloy wires versus traditional Au wire [32,33].

Alloy Wire Type	Advantages
Au-Pd system	Higher strength, better free air ball (FAB) morphology, lower IMCs growth
Au-Cu system	Higher strength, lower environmental requirements, lower IMCs growth
Au-In system	Better wire loop, lower break rate, higher bonding strength
Au-Cu-Ca system	Higher strength, better wire thinning effect, lower IMCs growth

2.3. Reliability of Au Bonding Wire

Bonding reliability is an important property of bonding wires. Compared with other types of bonding wires, Au bonding wire presents higher reliability, which is often used to connect aluminum terminals on semiconductor chips for signal transmission, and the Au-Al IMCs growth has a great impact on the bonding reliability according to many studies. Goh et al. [36] showed that the Au-Al IMCs formation makes the bonding stronger, which can improve the bonding reliability in an acidic environment. Liu et al. [37] investigated the microstructural evolution of Au-Al bonds, and the mechanical properties of Au-Al IMCs were computed by the first-principles calculations. The results showed that the Au–Al IMCs along with cracks grew during temperature increasing and aging time. Thermal exposure resulted in the transformation of the Au–Al phases associated with significant volumetric shrinkage leading to tensile stresses and promoted the growth of creep cavities that have a negative effect on the reliability of the Au–Al ball bonds [38]. In order to reduce the impact of Au-Al IMCs on the bonding reliability of electronics, scholars added trace elements such as Cu and Pd to make alloy wires to slow down the IMCs growth rate. Kim et al. [39] studied the effects on the interface reaction between Au-Pd alloy wire and Al pad, fitted the curve of IMCs thickness during thermal aging, and found that the Au-Pd alloy wire can reduce the IMCs growth rate by two orders of magnitude. Gam et al. [40] carried out a study on the growth mechanism of IMCs of Au-Pd alloy wire and Au-Cu alloy wire under high temperature storage test (HTST) at 150 °C. The results showed that both Cu and Pd can inhibit the Au-Al IMCs growth and pointed out that Cu presents a better improvement effect than Pd.

It is found that due to different atomic diffusion rates, Kirkendall voids (Figure 2 [41]) will be formed at the bonding interface, which often results in ball lifting from chip metallization and hence leads to catastrophic failure of the bond [42]. IMCs formation at the Au-Al interface caused Kirkendall voids, which resulted in interface resistance increase following non-constant Fickian law [43]. Du et al. [44] investigated the electromigration (EM) reliability and failure analysis of Au wire bonding process on Al substrate. The results showed that the interaction between the applied electric field and lattice defects increased the diffusing atomic migration and accelerated the IMCs formation. After aging at 130 °C for 200 h, Al depleted completely, and voids and cracks appeared inside the IMCs comprising Au_4Al and Au_8Al_3, as shown in Figure 3. Studies on the reliability of Au wires with different diameters showed that the Kirkendall voids were predominant in finer 0.6-mil and 0.8-mil Au ball bonds. No large-sized voids were observed in 1-mil and 2-mil wires, which further were completely absent in 3-mil wire ball bonds. It indicated that finer Au wires are more likely to produce Kirkendall voids defects and may no longer satisfy the requirements for a high density and fine pitch [45].

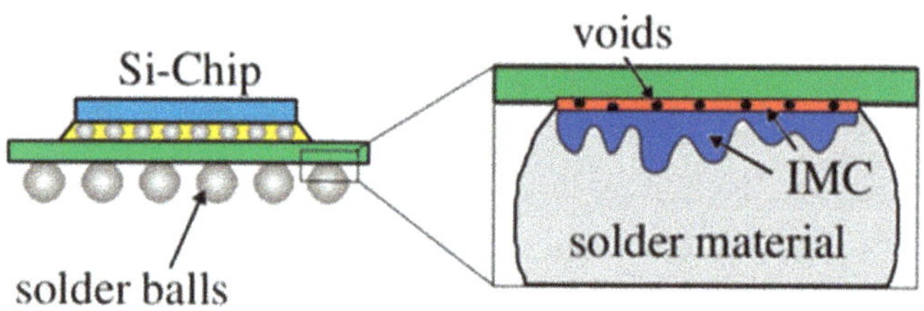

Figure 2. An example of Kirkendall voids in microelectronic packaging [41]. Copyright, 2020, Elsevier.

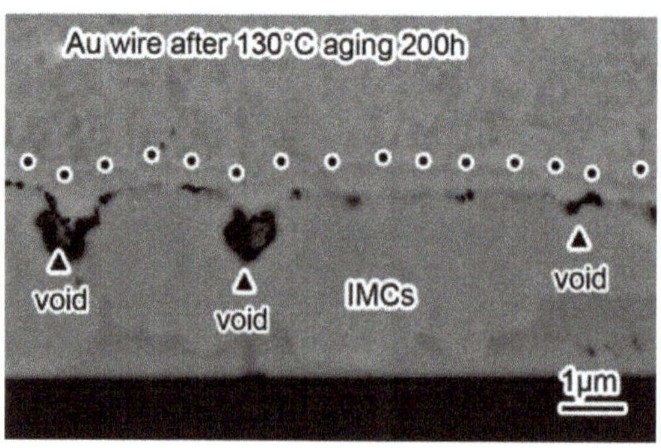

Figure 3. The cross-section images of bonding interface of Au wire after aging at 130 °C for 200 h [44]. Copyright, 2020, Springer Nature.

Design and optimization of experiments (DOE) methods and finite element analysis (FEA) have been widely utilized in both industry and academia, and they are also effective and low-cost ways to study the wire bonding process and its reliability [46]. Pan et al. [47] used FEA software to simulate the Au wire bonding process and obtained the stress/strain distribution of the FAB, thus providing theoretical support for predicting the bonding strength. 3D packaging is the development trend of the future market; Chen et al. [48] proposed an effective strategy for 3D stacked packaging and verified its feasibility by FEA, which will provide useful insights for the rapid development of wire interconnection of advanced 3D stacked microelectronics packaging and the reduction of development costs. The wire loop profile is also one of the important factors that affects the bonding reliability. At present, machine learning (ML) algorithm has received great attention in the process modeling, which provides vital guidance for establishing new wire loop models [49–51]. Hou et al. [52] proposed and demonstrated an efficient wire loop profile prediction model based on ML algorithm and FEA. It can complete the prediction of wire profile in seconds compared to the early experimental theoretical model and FEA model prediction, providing convenience for the research of wire bonding reliability.

With the continuous development towards multi-functional and highly integrated chips, microelectronic packaging is forced to develop in the direction of fine pitch, long distance, and high performance [53,54]. In consideration of wire sweep, bond crack, wire sag, and loop instability usually happening to ultra-fine Au bonding wire used for a high density and fine pitch bonding, many researchers and scientific institutions have paid more and more attention to new alternative bonding wires of low loop, high conductivity, and ultra-fine diameter such as Cu and Ag bonding wires to be better adapted to the demands of the microelectronics packaging market [55,56].

3. Cu Bonding Wire

Cu bonding wire, as one of the alternatives for Au bonding wire, includes bare Cu wire, Cu alloy wire, coated Cu wire, etc. Since the beginning of the 21st century, Cu-based wire has been widely used in IC packing, audio and video transmitters, active medical devices, and various electronic components; it is the key conductor material to ensure that the electrical system has steady power and signal transmission [57,58]. Compared with Au wire, Cu wire has lower cost, better electro-thermal performance, higher pull strength, and loop stability allowing for a reduced wire diameter to accommodate smaller pad sizes, and the growth rate of Cu-Al IMCs is smaller than one fifth of that of Au-Al IMCs, which greatly improves the chip frequency and bonding reliability [59–61].

3.1. Bare Cu Wire

Bare Cu wire as the conventional product of the Cu bonding wire has higher electrical conductivity, higher stiffness, and lower IMCs growing speed than Au wire, which contributes to less heat generation and void formation, better loop profile and reliability, and is more suitable for fine and ultra-fine pitch bonding [62]. It was reported that soft Cu wire slightly improves the bondability [63]. Some researchers confirmed that bare Cu wire has good reliability at an elevated temperature in high temperature storage testing [64,65] and dry atmosphere [66]. Experimental results showed that Cu bonds present higher tensile strength and shear strength than Au bonds with the same wire diameter [13,67]. Furthermore, Cu bonds potentially have higher durability than Au bonds due to the lower interdiffusion rate between the Cu and Al [68].

Although bare Cu wire has some advantages over Au wire, it still has many hurdles and cannot be used directly in industry [69]. Bare Cu wire is easy to be oxidized in air, and additional cost of forming gas, a mixture of 95% N_2 and 5% H_2, must be considered [70]. The higher hardness and stiffness of Cu wire over Au wire needs higher bonding force and more ultrasonic energy, which can damage the substrate, form die cratering, and induce pad peeling, pad crack, and Al splash, as shown in Figure 4 [71,72]. Wu et al. [73] showed that the critical factor leading to cracks in Cu wire bonding is that the lead frame flag floating on the thermal insert is caused by the shallower lead frame down-set or foreign matter on the thermal insert. To deal with the hurdles in Cu wire bonding, Li et al. [74]

increased pad thickness to reduce the impact and rebound of the Cu bonding process and thus improved the shear strength of Cu wire bonding. A new capillary was developed to improve stitch bondability, and satisfactory results were confirmed though stitch pull and ball shear tests [70]. It was found that high-purity Cu wire can improve its basic performance and bonding reliability. However, the high purity of Cu wire will give rise to its high cost, which further limits the use of bare Cu wire in microelectronic packaging. To expand the market of Cu bonding wire, many researchers dope trace elements to the Cu matrix to make alloy wires or coat other elements on its surface to make coated Cu wires [16,75,76].

(a) (b) (c)

Figure 4. Schematic diagram of pad defects during bare Cu wire bonding: (**a**) pad peeling; (**b**) pad crack; (**c**) Al splash [72]. Copyright, 2015, IEEE.

3.2. Cu Alloy Wire

Unlike bare Cu wire, Cu alloy wire is produced with a Cu base and trace elements such as alkali metals, alkaline earth metals, transition metals, and rare earth elements in varying concentrations that give it desired properties [77]. A patent revealed that Cu alloy wire added with Li 0.008–1.0 wt% and Ce 0.3–0.5 wt% had good corrosion resistance and short HAZ length that satisfied the requirements of high density and multi-layered packaging [78]. Fang et al. confirmed that Ca can improve the breaking resistance of Cu alloy wire and reduce its loop height. Meanwhile, Ce or Ti can improve the oxidation resistance and bondability of Cu alloy wire [79]. Yang et al. [80] drew the Cu and Cu-Fe alloy wires at room temperature, and the properties were characterized and analyzed. The results show that the strength and ductility of Cu-Fe wires are higher than those of Cu wire under the same drawing strain. Cu alloy wire with La 0.0008–0.002 wt%, Ce 0.001–0.003 wt%, Ca 0.002–0.004 wt%, and Cu 99.99–99.995 wt% had low hardness, low loop height, and good oxidation resistance that enabled it to satisfy the developing demands of high performance, multifunction, miniaturization, and portability of electronic packaging [81]. Another patent revealed that the adding of In, B, Bi, Ge, and Si to Cu matrix can increase the recrystallization temperature of Cu alloy wire and reduce its grain boundary motion rate that protected its grain boundary from cracking and reduced the damage to chip [82].

Particularly, it is well known that Cu-Ag alloys have excellent combination properties of high strength and high conductivity, and Ag has less influence on the conductivity of Cu alloys [83–85], which makes them the ideal materials for preparing fine bonding wires [86,87]. Continuous columnar-grained Cu-Ag alloy wires with identical diameter 0.3 mm but different Ag contents (6 wt%, 12 wt%, 24 wt%) were obtained at room temperature without intermediate annealing; the study results showed that incremental Ag contents contribute to significant improvement in tensile strength, while there is little variation in the elongation and the conductivity of the alloys, as shown in Figure 5 [88]. Xie et al. investigated the strengthening mechanisms of cold-rolled pure Cu and Cu-Ag alloys; the results verified that Ag addition increases the limiting concentration of dislocations and subgrain boundaries in the Cu solid solution, which leads to higher strength in the cold-rolled Cu-Ag alloys than that in pure Cu. Moreover, higher Ag content gives rise to higher strength of Cu-Ag alloys [89]. Zhu et al. [84] prepared a Φ40 μm Cu-4wt%Ag alloy wire with high strength and high electrical conductivity by continuously directional solidification and cold drawing. The main strengthening mechanism was grain refinement strengthening according to the calculation results of the strengthening model. The amount of trace elements added to Cu base should be controlled strictly in accordance with the bonding requirements; otherwise, it will have negative effects on the bonding properties of Cu alloy wire. Simultaneously, it was found that various degrees of cracks and fractures seem to happen to Cu alloy wires easily during the drawing process, which causes their low production efficiency and limits their applications in microelectronic packaging [90].

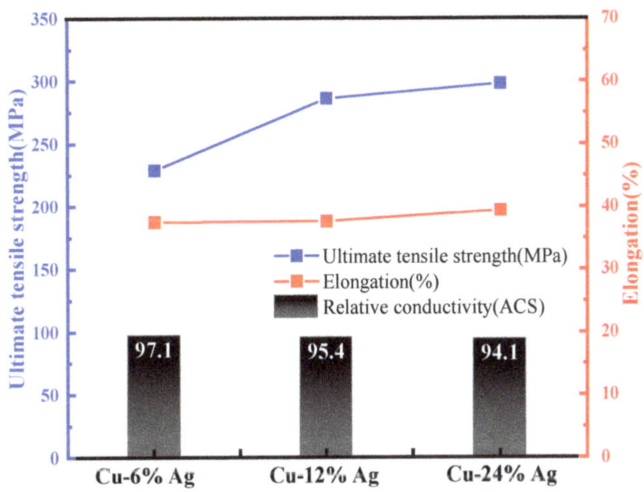

Figure 5. Mechanical-electrical properties of directionally solidified Cu-Ag alloys [88]. Copyright, 2018, Elsevier.

3.3. Coated Cu Wire

Coating the Cu wire with a pure metal layer which oxidizes slowly at processing temperature can extend shelf life and improve the corrosion resistance, bondability, and reliability of Cu bonding wire [91,92]. The coated pure metal layer comprises of noble metal elements such as Au, Ag, Pt, Pd and other metal elements of high corrosion resistance such as Ni, Co, Cr, and Ti. Among them, Pd shows stable performance and high corrosion resistance under long-term exposure to high temperature and high humidity conditions, and has good ductility, plasticity, and bending capacity, which adds an oxide-free and easily bondable surface for Cu wire [93,94]. Pd coating has excellent adhesion to Cu wire; round and stable FAB can typically be formed in a nitrogen atmosphere for Pd-coated Cu (PCC) wire [92,95]. PCC wire generally presents a much larger stitch bond window and more robust second bond, and shows better performance on Al pads in humidity stress and oxidation environments compared with bare Cu wire. It gains market share very fast as it can reduce chloride corrosion risk and resolve the limitation of bare Cu wire, especially on oxidation issues [96,97].

A PCC wire with Pd 1.35–8.19 wt% was invented by Zheng et al. that had long shelf life, high mechanical strength, and high oxidation resistance. It was beneficial to reduce the wire diameter and bonding pitch, and hence made it more applicable to high-density and multi-pin integrated circuit packaging [98]. Fang developed a new type of PCC wire with Ca, Mg, Al, and Sn added though the drawing process without intermediate annealing, which has preferential price, good plastic deformation capacity, and high pulling strength and reliability [99]. However, owing to the different mechanical properties of Cu and Pd, Pd coating is prone to flake off and bump during the drawing and annealing processes of

PCC wire, as shown in Figure 6, which makes the Cu matrix become oxidized in the air, resulting in reliability decrease of the bonding wire [100].

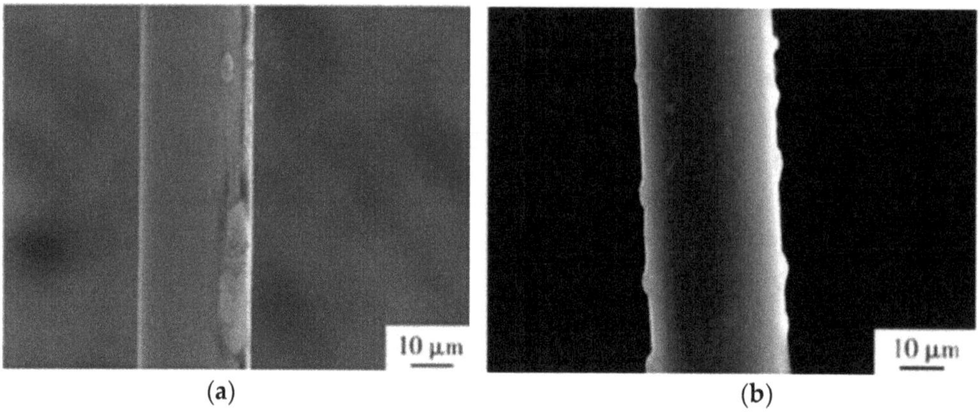

Figure 6. Surface morphology of Pd-coated Cu wire: (a) the coating flakes off; (b) the coating bumps [100]. Reproduced with permission from Ref. [100]; published by Heat Treatment of Metals, 2016.

3.4. Reliability of Cu Bonding Wire

Long-term reliability of Cu bonding wire is a major concern in replacing Au bonding wire. Cu-Al IMCs' phase and growth behavior affected the Cu wire bond reliability [101,102]. It was found that the growth rate of Cu-Al IMCs is much lower than that of Au-Al IMCs due to the larger atomic size difference and smaller electronegativity difference between Cu and Al than that between Au and Al, which leads to bonding failure of Au wire bonding due to Kirkendall voids (Figure 7a), while thin Cu-Al IMCs micro-crack (Figure 7b) [103], the results showed that Cu wire bonding has higher reliability than Au wire bonding. [65,104,105]. Diffusion-controlled IMCs of CuAl and $CuAl_2$ formed in the interface of the Cu-Al bonds after isothermal annealing by micro-X-ray diffraction [106]. Guo et al. [107] verified that $CuAl_2$ phase emerged first on the Cu-Al interface, which was consistent with that in the Al-Cu system studied by Xu et al. [108]. The sequence of IMP transformations during isothermal annealing from 175 °C to 250 °C was investigated by high resolution transmission electron microscopy (HRTEM). $CuAl_2$ and Cu_9Al_4 grew simultaneously, and the latter as a second layer (Figure 8) was the terminal product when the Al pad was completely consumed [105]. Experimental results related to interfacial structure development during the annealing process indicated that $CuAl_2$ is precipitated on the Cu-Al interface initially followed by Cu_9Al_4, and then the reaction phases of CuAl and Cu_4Al_3 formed at the bond interface [109,110]. The growth mechanism of Cu-Al IMCs was investigated by Yang et al. during the annealing temperature from 50 °C to 70 °C based on the in situ HRTEM, and the results showed that Cu-Al IMCs near and apart from the Cu layer were Cu_9Al_4 and $CuAl_2$, respectively. An accurate growth equation of Cu-Al IMCs was also obtained according to the in situ experimental results, which provides theoretical reference for the study of Cu-Al bond reliability [92]. Moreover, the presence of Pd at the bond interface can slow down the Cu-Al IMCs growth and preserve the bond strength during thermal annealing [111].

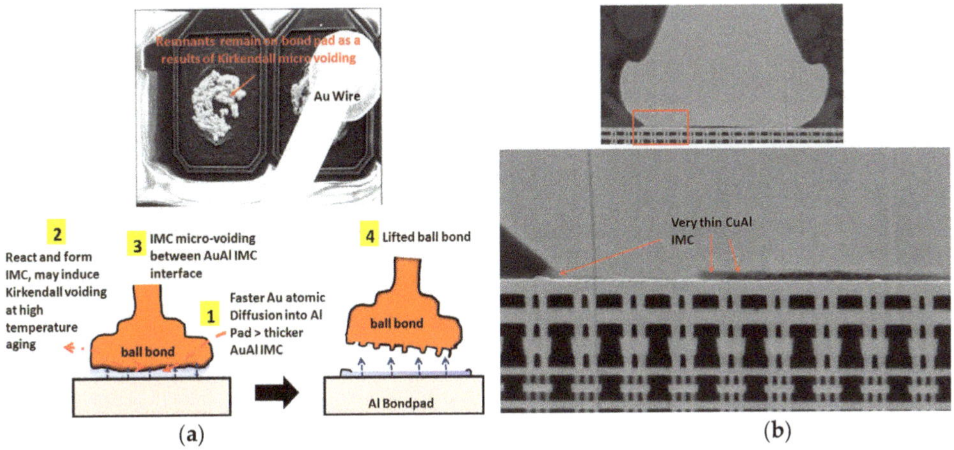

Figure 7. (**a**) Proposed failure mechanism of AuAl Kirkendall micro-voiding and caused lifted ball bond (**b**) SEM images show very thin CuAl IMC formation on bonded stage of Cu wirebond package prior to reliability stress. No microcracking beneath PdCu ball bond. [103]. Copyright, 2013, ASME Digital Collection.

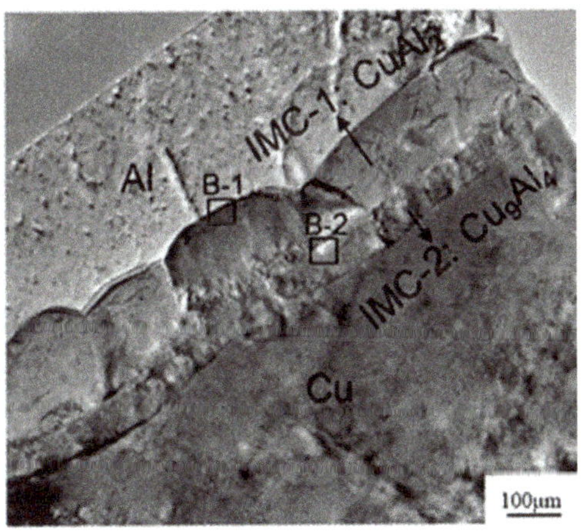

Figure 8. A typical TEM image of the Cu-Al interface after annealing at 175 °C for 25 h [105]. Copyright, 2011, Elsevier.

Various papers reported that compared with Au-Al bonds, lower Cu-Al IMCs' growth rate leads to less heat generation, lower electrical contact resistance, and better reliability and device performance [112]. Three bond pad metallizations (Al, AlCu, and AlCuSi) were used to compare Cu and Au wire bonding reliability under tests of operating life, temperature cycling, high temperature storage (HTS), thermal shock, pressure temperature humidity under bias (PTHB), temperature humidity under bias (THB), and Humidity Accelerated Stress Test (HAST), and the results showed that Cu wire bonding reliability is at least equal to the conventional Au wire bonding reliability [93]. However, the low bond strength and poor bondability at second bonds caused by surface oxidation of Cu wire limit its application in microelectronic packaging with high integration. To solve the above problems, Yuan et al. developed a Cu alloy bonding wire with Cu 99.75–99.96 wt%, W 0.01–0.1 wt%, Ag 0.01–0.03 wt%, Sc 0.01–0.02 wt%, Ti 0.001–0.03 wt%, Cr 0.001–0.03 wt%, and Fe 0.001–0.02 wt%, which has outstanding oxidation resistance and corrosion resistance, excellent plasticity and ballability, high electrical conductivity and thermal conductivity, and high strength and bonding reliability that enable it to meet the requirements of electronic packaging on high performance, multifunction, miniaturization, and low cost [113]. Murali et al. [114] added element of corrosion resistance to Cu to prepare a new alloyed Cu wire, which has homogeneous FAB formation and similar first and second bond bondability to bare Cu wire, and exhibits better electrochemical corrosion resistance and slower interfacial diffusion than that of bare Cu wire as well as higher bonding reliability than that of Au wire after HTST at 175 °C for 2000 h. Krimori et al. revealed that Cu bonding wire with oxidation-resistant metal coating Pd presents good bondability and reliability, sufficient to replace Au bonding wire (Table 2) as well as having much lower production cost than that of Au bonding wire [115,116]. In addition, electronic flame off (EFO) current has a strong influence on Pd distribution in the FAB [117], which has an effect on Cu–Al IMCs growth and thus affects bond strength and reliability [111]. Other literature also reported that the PCC wire improves bonding reliability under high temperature and high humidity and stressed environment due to the adjustment of controls for Cu-Al interdiffusion and specific IMCs formation [66,118,119].

Table 2. Reliability test results: deteriorated ratios after tests [115]. Copyright, 2006, IEEE.

Test	Duration	Failures by Wire Type		
		Pd-Coated Cu wire	Au Wire	Cu Wire
Temperature Cycle (−65 to −150 °C)	1000 cycles	0%	0%	91%
Temperature humidity bias (85 °C, 85%RH, 10V)	1000 h	0%	0%	3%
Pressure cooker test (PCT) (121 °C, 100%RH, 2 atom)	125 h	0%	0%	95%
Solder reflow (260 °C)	3 times	0%	0%	0%

The bonding wire used in the automotive industry is expected to change from Au wire to Cu wire in the coming years [120], and the AEC have published a new document which specifies the minimum requirements for qualification of Cu wire interconnections used in automotive electronics applications [121]. PCC wire has been widely used in large scale integration circuit (LSI) packages, but conventional PCC wire is difficult to achieve the target of automotive application. Eto et al. [122] developed a new PCC wire by adding elements in Cu core, which improved bond strength even under severe reliability test conditions. The bond pull test is the most common way to evaluate the bond strength in the wire bonding. Sun et al. [123] developed a sensor prototype for performance evaluation and applied it in a wire bonding pull test to evaluate the bond strength and provide support for wire bonding reliability research.

To sum up, Cu bonding wire has offered many benefits over the Au bonding wire such as high electrical conductivity, long wire spans without sagging, heat sinking capabilities, and low cost [124]. But the use of Cu bonding wire also has some drawbacks, mainly due to its hardness and susceptibility of Cu-Al IMCs to corrosion. Ag bonding wire, as a potential alternative wire, has better electrical and thermal conductivity than Au and Cu bonding wires as well as a relatively moderate price [125], and has been successfully bonded in die-to-die, overhanging die, DRAM stacked die [126], and multi die serial bonding in light emitting diodes (LEDs) [127].

4. Ag Bonding Wire

The price of Au has soared year by year; as an alternative to Au bonding wire, Cu bonding wire has been driven by its cost effectiveness, but it has its own concerns, such as sensitivity to corrosion and propensity to oxidize [128]. Since Ag has higher thermal and electrical conductivity than Au and Cu, Ag bonding wire can be chosen as an alternative material that can deliver higher reliability, lower cost, and better manufacturability than

Au and Cu bonding wire [129–131], and Ag bonding wire can be used in Au wire bonders of the same type [132].

4.1. Coated Ag Wire

Like Cu, Ag is easy to become corroded in sulfurated and oxidative environments [133]. Ag wires coated with Au or Pd were studied to improve the corrosion resistance and additional properties of pure silver, and the forms of coatings include single coating and composite coating [21]. Tseng et al. [134] showed that Au-coated Ag wire (ACA) is a promising material that can enhance bonding strength and the surface performance. The cross-sectional structure of the core material of ACA studied by Kang et al. [135] was equiaxed, which had small grain size and a large number of lath-like annealing twins, as shown in Figure 9. In terms of mechanical properties, electrical properties, and bonding properties, the test results were better than those of Au wire and pure Ag wire. Tanna et al. [129] produced a Pd coated Ag bonding wire (PCS) by a new method; the diameters of FABs produced with PCS in air are found to be more consistent than those of Au wire under the same test conditions. Further, the stitch pull force, neck pull force, and ball shear force of PCS were all higher than those of Au wire. Patent [136] developed a three-layer composed of Au, Pd, and Pt from the inside to the outside coated bonding wire, whose matrix was Ag-Au-Pd alloy. It has strong oxidation and sulfidation resistance, high bonding stability, and excellent effect of inhibiting Ag^{+} migration. There are still some shortcomings to be solved about the coated Ag wire, such as uneven coating, coating flaking off, and broken wire, which will affect the performance of the bonding wire and reduce the reliability of the product. Moreover, the usage of strong acids, strong bases, and organic solvents in the electroplating process will produce harmful substances and cause environmental pollution [137,138]. At present, there are few mature coated Ag-based bonding wires used in the market.

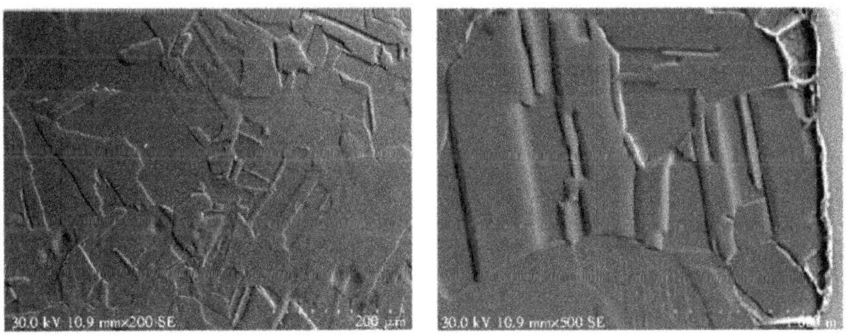

(**a**) (**b**)

Figure 9. SEM image of cross section of Au-coated Ag wire, (**a**) the magnification is 200 times and (**b**) the magnification is 500 times [135]. Reproduced with permission from Ref. [135]; published by Semiconductor Technology, 2018.

4.2. Ag Alloy Wire

The elements alloyed with Ag include Au, Pd, Pt, Ru, Zn, etc. The addition of Au can improve the strength, oxidation resistance, high-temperature stability of the Ag alloy wires, and enlarge the bondability window. Fan et al. [139] found that with the increases of Au content, the HAZ of the Ag alloy wire decreases, the tension and shearing force of the FAB increase, and the best FAB morphology can be obtained when the Au content is 5%. Kuo et al. [140] showed that Ag alloy wires exhibit superior tensile strength and hardness but low conductivity and elongation as Au content increases, because relatively high Au solute atoms and low-angle grain boundary density cause strengthening and electron scattering. Therefore, adjusting the Au content in Ag alloy wires effectively can optimize its mechanical properties and conductivity. Adding Au and Pd into Ag wire increases its corrosion resistance and improves its oxidation behavior by reducing the metal surface activity [141,142]. Tsai et al. [143] evaluated Ag alloy wires with various Au and Pd contents; the results indicated that the breaking load of the various wire specimens increases with the Pd and Au contents, as shown in Figure 10. Further, it was determined that the ternary Ag-8Au-3Pd alloy wire with high strength, corrosion resistance, and reliability is an ideal substitute for the traditional Au wire, while the Ag-3Pd and Ag-4Pd are cost-friendly bonding wires for high-frequency integrated circuit devices. Cao et al. [144,145] studied the effect of heat treatment temperature on the properties and microstructure of Ag–4Pd alloy wires. The results showed that the elongation and tensile strength of bonding wire increase and decrease, respectively, as heat treatment temperature increases, as shown in Figure 11. Annealing twins appear in the bonding wire in the process of heat treatment, and the nucleation modes mainly include twinning nucleation and sub-crystalline annexation. Chuang et al. [146,147] showed that an innovative Ag-8Au-3Pd alloy wire produced by appropriate drawing and annealing processes has more annealing twins, finer grains, and higher durability against EM than traditional Ag-8Au-3Pd alloy wire, as shown in Figure 12, which possesses high thermal stability during high temperature exposure. The annealing twins in the innovative wire simultaneously increase the ductility and tensile strength with aging time, yet the electrical conductivity remains almost unchanged. Moreover, the innovative wire exhibits negligible grain growth after prolonged air storage at 600 °C for up to 180 min, whereas the grain size of traditional Au and Cu wires grow obviously under the same conditions, which means that annealing twins may have the potential to improve material performance [148].

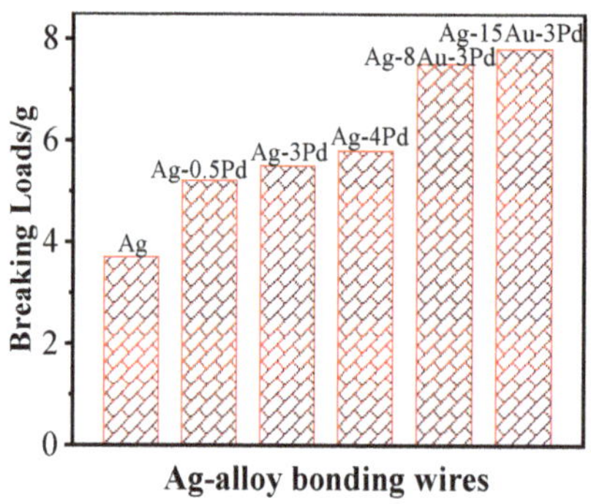

Figure 10. Fracture load of various Ag alloy wires [143]. Copyright, 2016, IEEE.

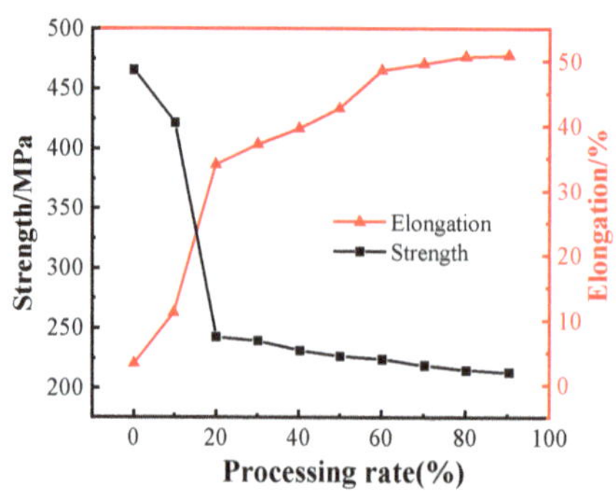

Figure 11. Variation curve of tensile strength and elongation of Ag-4Pd wire during cold deformation [145]. Reproduced with permission from Ref. [145]; published by Journal of Mechanical Engineering, 2016.

Figure 12. Grain structure in the longitudinal cross-sections of the original bonding wires: (**a**) annealed-twinned Ag-8Au-3Pd wire; (**b**) traditional Ag-8Au-3Pd wire [147]. Copyright, 2012, Springer Nature.

Recently, ultrasonic ribbon bonding has received more and more attention in power electronic packaging compared with traditional wire bonding. The high current density tolerance and excellent heat dissipation of ultrasonic ribbon bonding, which features the flexibility of wire bonding and wider bonding parameter window, have driven the adoption of it to meet the requirements of high reliability packaging [149–151]. Based on the successful experience of replacing Au with Ag wires, Ag and Ag alloy ribbons are foreseeable options for high-power ICs. Chen et al. [151] studied the microstructure evolution of Ag-4Pd ribbon under different annealing conditions by electron backscatter diffraction (EBSD) technique, and discussed its mechanical properties, which provided a theoretical basis for the application of high-power IC modules. Chen et al. [152] further studied the influence of bonding parameters on the bonding performances of Ag-4Pd ribbon. The results showed that increasing the bonding power can promote the bonding surface, and the research has important guiding significance for providing feasible solutions for electronic packaging.

4.3. Other Ag Alloy Wires

In order to develop Ag alloy wire that is more in line with production requirements, some scholars have tried adding other trace elements, such as Pt, Zn, and La, to the Ag matrix to make further progress. Pt, as noble metal with similar properties to Pd, can form a continuous solid solution with Ag, whose addition amount is generally 5 wt% [35]. Ag alloy wires containing rare earth elements have excellent anti-oxidation and mechanical properties [153,154]; the additions of Si, Zn, etc., could reduce the oxygen content of Ag bonding wire, improve its wettability, and make it easier to be drawn [34]. Cao et al. [155] added Zn 0.76 wt% to Ag-1.7Pd alloy wire; it was found that the coefficient of heat conductivity reduces by 16%, and the HAZ length decreases by 25%, as shown in Figure 13. Hsueh et al. [154] doped La into Ag to form Ag–La alloy; the results showed that adding La can improve its anti-oxidation capacity, reduce the FAB diameter, and increase the hardness of the matrix, which make it useful in the electronic packaging industry.

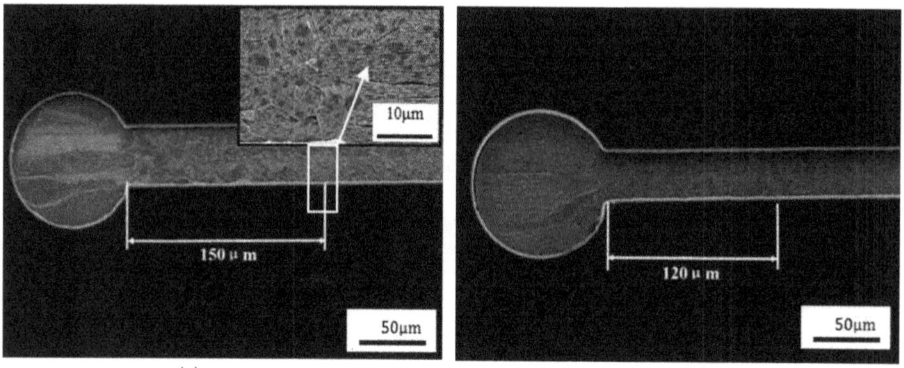

(a) (b)

Figure 13. HAZ length of Ag alloy wire (**a**) Ag-1.7Pd wire; (**b**) Ag-1.7Pd-0.75Zn wire [155]. Reproduced with permission from Jun Cao et al.; published by IEEE, 2015.

4.4. The Reliability of Ag Bonding Wire

Up to now, Ag bonding wire has not been widely used in microelectronic packaging industry because of its low bonding reliability, which is affected by FAB morphology, Ag^{+} migration, IMCs, doping element types, contents, and so on.

FAB morphology has a great influence on the reliability of Ag wire bonding; the poor FAB morphology, containing nodes and voids on FAB, can cause flaws at the bonding interface [130]. Cao et al. [156] studied the effect of Au layer thicknesses on FAB morphology of ACA, and obtained excellent FAB morphology when the Au coating thickness was 108 nm, as shown in Figure 14. They also proved that the FAB size of the Ag-10Au-3.6Pd alloy bonding wire became larger and larger with the increase of EFO time or EFO current, and an optimal FAB formed when the EFO current is 0.030A and the EFO time is 0.8 ms [157]. The experimental results provided that a good FAB morphology can be obtained in high current and short-time EFO process, which is beneficial to improve the bondability [158]. Therefore, setting appropriate bonding parameters is vital for FAB morphology, and it will directly have an effect on electronics reliability.

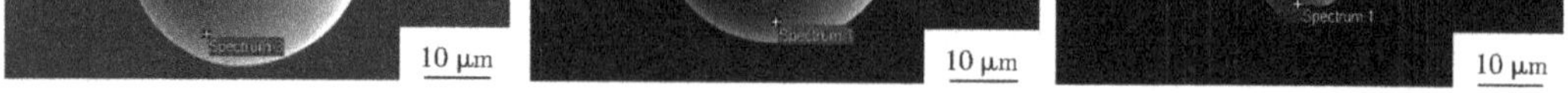

(a) (b) (c)

Figure 14. Morphology of FAB for Au-coated Ag bonding wires with (**a**) 48.7 nm, (**b**) 108 nm, and (**c**) 174.7 nm Au coating thickness [156]. Reproduced with permission from Ref. [156]; published by Materials Science and Technology, 2018.

For decades, it has been known that Ag^{+} migration can cause short circuits and failure in microelectronic interconnects [159]. In order to inhibit Ag^{+} migration to improve bonding reliability effectively, the experimental results showed that the addition of 1.5–4.5 wt% Pd can decrease the Ag^{+} migration rate [160], and the effect is further enhanced due to the

Au addition [161,162]. Cho et al. [163] confirmed that a PdO layer forms on the surface Pd-doped Ag wire, which acts as a barrier against Ag^+ migration and diffusion to prevent Ag from dissolving in the surrounding water, as shown in Figure 15. Cai et al. [164] showed that a large number of annealing twins form in the Ag-4Pd alloy wire by adding Pd, which has positive effects on reducing the Ag^+ migration rate and improving the reliability of the bonding wire. The research results of Guo et al. [165] and Chuang et al. [147] indicated that the Ag-8Au-3Pd alloy wire with annealing twins has significantly enhanced resistance to Ag^+ migration, and Guo et al. supposed that a metal cap coating on the wire surface may be a better choice to prevent Ag^+ migration. According to the above research, adding Pd to Ag alloy wires and preparing bonding wires with annealing twins can effectively inhibit the Ag^+ migration and improve the product reliability. Recently, due to the development of power electronics and high-frequency communication ICs (such as fifth generation mobile networks and electric automotive), concerns about circuit reliability have further increased. Chen et al. [166] used EBSD to study the microstructure evolution of Ag alloy wire, providing a new perspective for the study of Ag^+ migration.

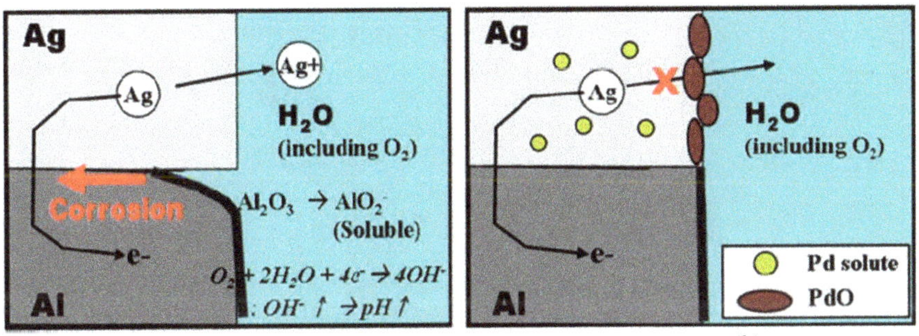

Figure 15. Mitigation mechanism of corrosion: (**a**) bare Ag wire; (**b**) Ag-3Pd alloy wire (Effect on Pd alloying-Ag migration model) [163]. Copyright, 2012, Springer Nature.

In addition, the growth of IMCs at the bonding interface has a great influence on the bonding reliability. A certain number of IMCs can improve the bonding strength, while due to the poor electro-thermal properties of IMCs, excessive IMCs will weaken the bonding strength and conductivity, and affect the lifetime of electronic products [167]. Hsu et al. [168] showed that high current density will increase the growth rate of IMCs, which will make IMCs form more quickly at anode side than that at cathode side, and the IMCs include Ag_3Al and Ag_2Al at the bonding interface. In the high temperature aging test, the Al pad completely disappears with the increase of temperature and aging time, then Ag_3Al begins to transform to Ag_2Al, and finally IMCs are all Ag_2Al [169]. The experimental results showed that with the increase of bias cycle, the growth of Ag_2Al at the bonding interface is responsible for the increase in the resistance of the bonding wire [170]. Chuang et al. [171] proved that even after prolonged high HTS at 150 °C for 500 h, IMCs thickness of Ag-8Au-3Pd is only around 1.7 μm, which has much higher reliability than that of the Au-Al IMCs. Long et al. [172] designed and created different surface textures on glass substrates. The results showed that the oxide removal efficiency is enhanced, the IMC growth is improved, and the bonding strength is several times higher than that on smooth surfaces. The above studies show that the quantities and types of IMCs are also the keys for the reliability. Up to now, most studies focus on Ag alloy wires, some of which have been used in LED and IC packages such as Ag-8Au-3Pd and Ag-4Pd wires [143]. However, uneven mixed or unreasonable chosen alloy components will lead to many defects during application, such as sharp FAB, golf-clubbed ball, and wave ball, which have negative effects on bonding reliability. Thus, it is necessary to make great efforts to carry on further study on the reliability of Ag bonding wire.

5. Comparisons

As reviewed in the aforementioned sections, Au bonding wire shows the best comprehensive performance, but has the highest price. Compared with Au bonding wire, Cu and Ag bonding wires, as potential alternative wires, have better cost effectiveness and higher electrical and thermal conductivity, which enable them to satisfy the requirements for a high density and fine pitch of microelectronic packaging and gain market share fast. The performance comparison of the three types of bonding wires is shown in Table 3. All the three types of bonding wires still face some technical challenges, and numerous researchers make great efforts to solve them as shown in Table 4.

Table 3. Basic properties of Au, Cu and Ag [17,173–175].

Properties	Au	Cu	Ag
Electrical Conductivity (% IACS)	73.4	103.1	108.4
Resistivity ($\times 10^{-9}$ Ω·m)	23.5	16.7	14.7
Thermal Conductivity (W/m·K)	317.9	398.0	428.0
Thermal Expansion Coefficient (μm/m·K)	14.2	16.7	19.0
Tensile Strength (MPa)	103.0	209.0	125.0
Yield Strength (MPa)	30.0–40.0	33.3	35.0
Elastic Modulus (GPa)	78.0	128.0	71.0
Brinell Hardness (HB)	18.0	37.0	25.0
Metal Activity		Cu > Ag > Au	

Table 4. Challenges and solutions for bonding wires.

Type	Technical Challenges	Solution
Au	Poor loop formability and easy to collapse during wire bonding.	Add alloying elements such as Cu and Pd; use Cu bonding wire to replace it.
	Easy to generate Kirkendall voids under long-term high temperature storage life (HTSL) tests.	Add trace alloying elements to slow down the diffusion rate of Au atoms.
Cu	Easy to be oxidized and corroded.	Chose alloying and plating treatment; use inert shielding gas.
	High hardness, causing defects such as pad peeling and cratering.	Use softer copper wire; increase pad thickness; set dummy microvias beneath pad metallization to stabilize and strengthen pad structure.
	Low hardness and strength of the HAZ, resulting in wire breakage near the FAB.	Adjust the EFO parameters.
	Short circuit and tail defects.	Use Pd-coated Cu wire instead of bare Cu wire.
	Cu-Al IMC corrosion, causing microcracks.	Use low halogen of molding compound and solder resist in the pad.
	$Al(OH)_3$ forms during bonding process, galvanic and pitting corrosion occur with the presence of chloride halides in sodium chloride solutions.	Control humidity and the temperature of production workshop [176].
Ag	Easy to be oxidized and vulcanized.	Coat Au, Pd, etc. on bare Ag wire; add alloying elements to make Ag alloy wire.
	Ag^+ migration.	Add Pd to inhibit Ag^+ migration.
	Low PCT reliability of bare Ag wire.	Increase Pd content reasonably.
	Complex bonding process with narrow process window.	Optimize the bonding process; add alloying elements such as Au and Pd.
	Short tailing during wire bonding.	Reduce the second bonding force by 10%.
	Poor FAB size repeatability and concentricity of Ag alloy wire.	Adopt lower FAB flow rates and EFO currents [72].
	FAB surface deformation and defects.	Increase Pd content.

6. Summaries

Wire bonding is a promising and critical interconnection technology with low cost, large-span ratio, and high-density packaging capability, which has been applied in microelectronic packaging for more than 70 years. Due to the rapid bonding process, the evolution of bonding interface is difficult to observe, and the bonding mechanism cannot be clearly described. Au bonding wire is the earliest and most widely used bonding wire, but due to the limit of electro-thermal performance and high price, its market share is gradually decreasing; currently, it is mainly used in high-end products applied in aerospace and other fields. The application of Cu bonding wire came out in the 1990s, and its market share has increased rapidly in recent years, especially Pd-coated Cu wire, which has become one of the most used bonding wires in low-pin-count semiconductor packaging, Radio Frequency (RF) device packaging, micro-electromechanical systems, and so on, but some reliability problems still need to be solved. Ag bonding wire began to appear in the market in the 2010s, and it has been widely used in IC packaging and LED packaging for many years. Because of its excellent usability, scholars believe that it will replace Au bonding wire applied in high-end products. In addition, multilayer wire bonding is the development trend of the future market, and the design of wire loop profile is also the focus of research.

In general, the research tasks of bonding wires in the future may focus on the following aspects:

(1) It is difficult to observe the evolution of the bonding interface during the wire bonding process, and the wire bonding mechanism is still unclear. Therefore, developing new research techniques can play important roles in studies on the wire bonding mechanism.

(2) The electroplating and electroless plating processes currently used in the preparation of coating bonding wires will produce harmful substances that may pollute the environment, so it is necessary to develop new environmentally friendly and resource-efficient surface treatment technology to prepare coating bonding wires.

(3) Microalloying can improve the mechanical properties of bonding wires, so appropriate multi-component microalloy doping process is of significance in the preparation of bonding wires, which is beneficial to increase the market share of Cu-based and Ag-based bonding wires.

(4) The bonding wires with high cost and poor performance have not been enough to satisfy the requirements of electronic packaging on high performance, multifunction, miniaturization, and low cost, so developing Cu-based or Ag-based bonding wires with cost effectiveness and excellent performance has become more and more popular for scholars and manufacturers.

(5) The wire loop profile has a great impact on the bonding reliability. The main challenge is to control the wire loop height in the ultra-thin packaging space. It is the future trend to study the wire profile models and the reliability of wire bonding by FEA and ML algorithms.

Author Contributions: Conceptualization, H.Z. and J.C.; methodology, H.Z.; validation, J.C., A.C. and B.A.; investigation, Y.Z.; resources, Y.Z., A.C. and B.A.; data curation, C.S., B.A. and C.L.; writing—original draft preparation, Y.Z., C.S., C.L. and A.C.; writing—review and editing, Y.Z., H.Z., A.C. and B.A.; supervision, H.Z.; project administration, H.Z.; funding acquisition, H.Z. and J.C. All authors have read and agreed to the published version of the manuscript.

Funding: This research was funded by Joint Funds of the National Natural Science Foundation of China, grant number U21A2051; Key Science and Technology Program of Henan Province, grant number 222102230019; The Fundamental Research Funds for the Universities of Henan Province, grant number NSFRF210102; Doctoral Fund of Henan Polytechnic University, grant number B2015-40.

Institutional Review Board Statement: Not applicable.

Informed Consent Statement: Not applicable.

Data Availability Statement: Not applicable.

Conflicts of Interest: The authors declare no conflict of interest.

References

1. Alim, M.A.; Abdullah, M.Z.; Aziz, M.S.A.; Kamarudin, R. Die attachment, wire bonding, and encapsulation process in LED packaging: A review. *Sens. Actuators A Phys.* **2021**, *329*, 112817. [CrossRef]
2. Hamid, K.A.; Badarisman, A.H.; Jalar, A.; Bakar, M.A. Investigation of integrated factors in the occurrence of copper wire bonding corrosion of semiconductor packages. *J. Phys. Conf. Ser.* **2022**, *2169*, 12016. [CrossRef]
3. Su, L.; Yu, X.N.; Li, K.; Pecht, M. Defect inspection of flip chip solder joints based on non-destructive methods: A review. *Microelectron. Reliab.* **2020**, *110*, 113657. [CrossRef]
4. Ershova, N.Y.; Reginya, S.A.; Lunkov, P.V.; Kirienko, D.A. Multi-chip assemblies combining wire bond and flip-chip package technologies. *IOP Conf. Ser. Mater. Sci. Eng.* **2020**, *862*, 52059. [CrossRef]
5. Liu, D.S.; Chen, C.Y.; Chao, Y.C. A Thermomechanical Study of the Electrical Resistance of Cu Lead Interconnections. *J. Electron. Mater.* **2006**, *35*, 958–965. [CrossRef]
6. Jaafar, N.B.; Choong, C.S. Establishment of Highly Dense Wire Bonding with Insulated Au Wire. In Proceedings of the 2022 IEEE 24th Electronics Packaging Technology Conference (EPTC), Singapore, 7–9 December 2022.
7. Gomes, J.; Mayer, M.; Lin, B. Development of a fast method for optimization of Au ball bond process. *Microelectron. Reliab.* **2015**, *55*, 602–607. [CrossRef]
8. Yu, C.M.; Lai, K.K.; Chen, K.S.; Chang, T.C. Process-Quality Evaluation for Wire Bonding with Multiple Gold Wires. *IEEE Access* **2020**, *8*, 106075–106082. [CrossRef]
9. Kumar, G.I.A.; Lambert, A.; Caperton, J.; Asokan, M.; Yi, W.; Chyan, O. Comparative Study of Chloride and Fluoride Induced Aluminum Pad Corrosion in Wire-Bonded Device Packaging Assembly. *Corros. Mater. Degrad.* **2021**, *2*, 447–460. [CrossRef]
10. Ding, K.W. Research on Bonding Reliability of Semiconductor Gold Wire. Master's Thesis, University of Chinese Academy of Sciences, Beijing, China, 2014.
11. Schneider-Ramelow, M.; Ehrhardt, C. The reliability of wire bonding using Ag and Al. *Microelectron. Reliab.* **2016**, *63*, 336–341. [CrossRef]
12. Chuang, T.H.; Lin, H.J.; Chuang, C.H.; Shiue, Y.Y.; Shieu, F.S.; Huang, Y.L.; Hsu, P.C.; Lee, J.D.; Tsai, H.H. Thermal stability of grain structure and material properties in an annealing twinned Ag–4Pd alloy wire. *J. Alloys Compd.* **2014**, *615*, 891–898. [CrossRef]
13. Gu, B.K.; Shen, S.N.; Li, H. Mechanism of microweld formation and breakage during Cu–Cu wire bonding investigated by molecular dynamics simulation. *Chin. Phys. B* **2022**, *31*, 016101. [CrossRef]
14. Zhang, H.; Guo, Z.N.; Li, Y.B.; Yuan, C.; Chen, Y.J. Research progress of fine copper wire bonding technology. In Proceedings of the 14th National Special Processing Academic Conference, Suzhou, China, 22–25 October 2011.
15. He, X.K.; Guo, L.B.; Gong, G.S.; Su, F.L.; Zhu, D.C. Effects of different inhibitor on antioxidation of copper bonding wire at room temperature. *J. Mater. Sci. Mater. Electron.* **2022**, *33*, 10561–10571. [CrossRef]
16. Motoki, E.; Noritoshi, A.; Takashi, Y.; Masaaki, S.; Shinji, F. Microstructural characterization of alloyed palladium coated copper wire under high temperature. *Microelectron. Reliab.* **2021**, *120*, 114125.
17. Manoharan, S.; Patel, C.; Mccluskey, P. Advancements in Silver Wire Bonding. In Proceedings of the ASME 2017 International Technical Conference and Exhibition on Packaging and Integration of Electronic and Photonic Microsystems, San Francisco, CA, USA, 29 August 2017.
18. Zhao, J. Discussion the characteristics of coating Pd copper wire bonding. *Electron. Packag.* **2012**, *12*, 36–41.
19. Li, X.W.; Cai, J.L.; Chen, L.; Lin, J.T. Analysis and comparison of performance of bonding wire in LED packaging. *China Light Light.* **2019**, *49*, 18–22.
20. Hsueh, H.W.; Hung, F.Y.; Lui, T.S.; Chen, L.H. Microstructure, electric flame-off characteristics and tensile properties of silver bonding wires. *Microelectron. Reliab.* **2011**, *51*, 2243–2249. [CrossRef]
21. Zhong, M.J.; Huang, F.X.; Yuan, H.G. Research progress on the copper and silver bonding wire materials. *Mater. Rep.* **2017**, *31*, 99–102.
22. Xie, Q.; Long, K.; Lu, D.N.; Li, D.W.; Zhang, Y.; Wang, J. Integrated Circuit Gold Wire Bonding Measurement via 3-D Point Cloud Deep Learning. *IEEE Trans. Ind. Electron.* **2022**, *69*, 11807–11815. [CrossRef]
23. Cao, J.; Wu, W.X. Research and Development Prospect of Ag Alloy Bonding Wire in Microelectronic Packaging. *Precious Met.* **2017**, *38*, 7–11.
24. Papadopoulos, C.; Villiger, T.; Steiner, S.; Prinz, J.; Morabito, C.; Cammarata, M.; Rodriguez, A.; Kwon, J.; Geilenkeuser, R.; Breuer, D.; et al. Gold wire bond study for automotive application. *Microelectron. Reliab.* **2020**, *114*, 113899. [CrossRef]
25. Xie, S.; Lin, P.R.; Yao, Q.B. Interface mechanical behavior of gold alloy wire bonding. *J. Phys. Conf. Ser.* **2021**, *1907*, 12021. [CrossRef]
26. März, B.; Graff, A.; Klengel, R.; Petzold, M. Interface microstructure effects in Au thermosonic ball bonding contacts by high reliability wire materials. *Microelectron. Reliab.* **2014**, *54*, 2000–2005. [CrossRef]
27. Humpston, G.; Jacobson, D.M. A new high strength gold bond wire. *Gold Bull.* **1992**, *25*, 132–145. [CrossRef]
28. Wang, S.P.; Zhao, J.Y.; Xu, B.Q.; Kong, L.X.; Jiang, W.L.; Yang, B. Theoretical research on vacuum separation of Au-Ag alloy. *T. Nonferr. Metal. Soc.* **2020**, *32*, 2719–2726. [CrossRef]
29. Zhu, J.G. Trend of Alloying Investigation for Gold Bonding Wires. *Rare Met.* **2002**, *23*, 57–61.
30. Ning, Y.T. Properties and applications of some gold alloys modified by rare earth additions. *Gold Bull.* **2005**, *38*, 3–8. [CrossRef]

31. Simons, C.; Schräpler, L.; Herklotz, G. Doped and low-alloyed gold bonding wires. *Gold Bull.* **2000**, *33*, 89–96. [CrossRef]
32. Hong, H.L.; Yang, C.H.; Wang, K.C.; Gao, H.; Yan, H.X. Composition analysis of AU–AG and AU–CU solid solution alloys by Friedel's periodic spherical oscillation model. *Mod. Phys. Lett. B* **2021**, *35*, 2141015. [CrossRef]
33. Kim, H.G.; Lee, T.W.; Jeong, E.K.; Kim, W.Y.; Lim, S.H. Effects of alloying elements on microstructure and thermal aging properties of Au bonding wire. *Microelectron. Reliab.* **2011**, *51*, 2250–2256. [CrossRef]
34. Lin, L.; Zang, X.D. Introduction of a Corrosion-resistant High-reliability Silver AlloyBonding Wire in Package. *Electron. Packag.* **2014**, *14*, 9–13.
35. Tang, R.; Yang, X.L.; Wu, B.A.; Tang, H.Y.; Zhang, C. Analysis and prospect on alloy system and compositions of gold–silver alloy bonding wire from the perspective of patent citations. *Electron. Compon. Mater.* **2019**, *38*, 8–13.
36. Goh, C.; Chong, W.; Lee, T.; Breach, C. Corrosion Study and Intermetallics Formation in Gold and Copper Wire Bonding in Microelectronics Packaging. *Crystals* **2013**, *3*, 391–404. [CrossRef]
37. Liu, P.; Cong, S.; Tan, X.W.; Lin, X.; Wu, P. The reliability assessment of Au–Al bonds using parallel gap resistance microwelding. *J. Mater. Sci. Mater. Electron.* **2020**, *31*, 6313–6320. [CrossRef]
38. Noolu, N.J.; Murdeshwar, N.M.; Ely, K.J.; Lippold, J.C.; Baeslack, W.A. Degradation and failure mechanisms in thermally exposed Au–Al ball bonds. *J. Mater. Res.* **2004**, *19*, 1374–1386. [CrossRef]
39. Kim, H.J.; Song, M.; Paik, K.; Moon, J.; Song, J. Investigation of interfacial phenomena of alloyed Au wire bonding. In Proceedings of the 2013 IEEE 15th Electronics Packaging Technology Conference (EPTC 2013), Singapore, 1 December 2013.
40. Gam, S.A.; Kim, H.J.; Cho, J.S.; Park, Y.J.; Moon, J.T.; Paik, K.W. Effects of Cu and Pd addition on Au bonding wire/Al pad interfacial reactions and bond reliability. *J. Electron. Mater.* **2006**, *35*, 2048–2055. [CrossRef]
41. Zhang, F.; Zhang, B.Y.; Chen, X.P.; Zhang, X.H.; Zhu, X.K.; Du, H.S. Computational simulation of voids formation and evolution in Kirkendall effect. *Physical A* **2020**, *554*, 124285. [CrossRef]
42. Schneider-Ramelow, M.; Schmitz, S.; Schuch, B.; Grübl, W. Kirkendall voiding in au ball bond interconnects on al chip metallization in the temperature range from 100–200 °C after optimized intermetallic coverage. In Proceedings of the 2009 European Microelectronics and Packaging Conference (EMPC), Rimini, Italy, 15–18 June 2009.
43. Ahmad, S.S.; Smith, S.C. Au/Al Wire Bond Interface Resistance Degradation Rate Simulations. *IEEE Trans. Device Mater. Reliab.* **2019**, *19*, 774–781. [CrossRef]
44. Du, Y.H.; Gao, L.Y.; Yu, D.Q.; Liu, Z.Q. Comparison and mechanism of electromigration reliability between Cu wire and Au wire bonding in molding state. *J. Mater. Sci. Mater. Electron.* **2020**, *31*, 2967–2975. [CrossRef]
45. Murali, S.; Srikanth, N.; Vath, C.J. Effect of wire size on the formation of intermetallics and Kirkendall voids on thermal aging of thermosonic wire bonds. *Mater. Lett.* **2004**, *58*, 3096–3101. [CrossRef]
46. Tang, J.Y.; Li, L.T.; Zhang, G.Q.; Zhang, J.; Liu, P. Finite element modeling and analysis of ultrasonic bonding process of thick aluminum wires for power electronic packaging. *Microelectron. Reliab.* **2022**, *139*, 114859. [CrossRef]
47. Pan, Y.B.; Zhu, F.L.; Lin, X.X.; Fan, J.J.; Liu, S. Comparison of ultrasonic wire bonding process between gold and copper by nonlinear structure analysis. *J. Adhes. Sci. Technol.* **2018**, *32*, 2007–2018. [CrossRef]
48. Chen, Y.; Ding, S.Q.; Long, J.Y.; Hou, M.X.; Chen, X.; Gao, J.; He, Y.B.; Wong, C.P. Rationally Designing the Trace of Wire Bonder Head for Large-Span-Ratio Wire Bonding in 3D Stacked Packaging. *IEEE Access* **2020**, *8*, 206571–206580. [CrossRef]
49. Huang, W.J.; Samanta, A.; Chen, Y.; Baek, S.; Shaw, S.K.; Ding, H.T. Machine learning model for understanding laser superhydrophobic surface functionalization. *J. Manuf. Process.* **2021**, *69*, 491–502. [CrossRef]
50. Nazir, Q.; Shao, C.H. Online tool condition monitoring for ultrasonic metal welding via sensor fusion and machine learning. *J. Manuf. Process.* **2021**, *62*, 806–816. [CrossRef]
51. Wang, P.; Yang, Y.R.; Moghaddam, N.S. Process modeling in laser powder bed fusion towards defect detection and quality control via machine learning: The state-of-the-art and research challenges. *J. Manuf. Process.* **2022**, *73*, 961–984. [CrossRef]
52. Hou, M.X.; Ou, Z.P.; Long, J.Y.; Ding, S.Q.; Wen, G.H.; Chen, Y.; Chen, X. Machine learning enables accurate wire loop profile prediction for advanced microelectronics packaging. *J. Manuf. Process.* **2022**, *84*, 394–402. [CrossRef]
53. Cao, J.; Zhang, J.C.; Wu, B.A.; Tang, H.Y.; Lv, C.C.; Song, K.X.; Yang, G.N.; Cui, C.Q.; Gao, Y.G. Study on Manufacturing Technology of Ag-8.5Au-3.5Pd Fine Alloy Wire. *Micromachines* **2021**, *12*, 938. [CrossRef]
54. Ding, U.T.; Cao, J.; Xu, G.J.; Kou, S.Z.; Hu, R. Research and Application of Copper Bonding Wire in Electronic Packaging. *Foundry Technol.* **2006**, *27*, 971–974.
55. Yuan, F.S. Current Situation and Development Trend of Bonding Wire Market. In Proceedings of the 2021 China Copper Processing Industry Annual Conference and China (Sanmenxia) Copper Industry High Quality Development Conference, Sanmenxia, China, 10 June 2021.
56. Xu, H.; Liu, C.; Silberschmidt, V.V.; Pramana, S.S.; White, T.J.; Chen, Z.; Acoff, V.L. Intermetallic phase transformations in Au–Al wire bonds. *Intermetallics* **2011**, *19*, 1808–1816. [CrossRef]
57. Li, Z.; Xiao, Z.; Jiang, Y.B.; Lei, Q.; Xie, J.X. Composition design, phase transition and fabrication of copper alloys with high strength and electrical conductivity. *Chin. J. Nonferrous Met.* **2019**, *29*, 2009–2049.
58. Song, K.X.; Zhou, Y.J.; Mi, X.J.; Xi, Z.; Cao, J.; Ding, Y.T.; Wu, B.A.; Feng, C.L.; Li, Z.; Chen, D.B.; et al. Preparation, microstructure and properties of copper based wire. *Chin. J. Nonferrous Met.* **2020**, *30*, 2845–2874.
59. Hamid, K.A.; Badarisman, A.H.; Jalar, A.; Bakar, M.A. Effects of electrolyte towards copper wire metallurgical interconnection in semiconductor. *J. Phys. Conf. Ser.* **2022**, *2169*, 012013. [CrossRef]

60. Lu, K.; Ren, C.L.; Gao, N.Y.; Ding, R.Z. The Process and Reliability Researches of Copper Wire Bonding. *Electron. Packag.* **2010**, *10*, 1–6.
61. Gan, C.L.; Ng, E.K.; Chan, B.L.; Classe, F.C.; Kwuanjai, T.; Hashim, U. Wearout Reliability and Intermetallic Compound Diffusion Kinetics of Au and PdCu Wires Used in Nanoscale Device Packaging. *J. Nanomater.* **2013**, *2013*, 1–9. [CrossRef]
62. Zhong, Z.W. Wire bonding using copper wire. *Microelectron. Int.* **2009**, *26*, 10–16. [CrossRef]
63. Kam, J.; Meng, H.; Calipito, D.; Dittmer, K. Materials characteristics of soft copper wires designed for advance application. In Proceedings of the SemiCon Singapore 2007, Singapore, 15 May 2007.
64. Wulff, F.W.; Breach, C.D.; Stephan, D.; Dittmer, K.; Garnier, M. Further Characterisation of Intermetallic Growth in Copper and Gold Ball Bonds on Aluminium Metallisation. In Proceedings of the Proc SEMICON®Singapore 2005, Singapore, 6 May 2005.
65. Kim, H.; Lee, J.Y.; Paik, K.; Koh, K.; Won, J.; Choe, S.; Lee, J.; Moon, J.; Park, Y. Effects of Cu/Al intermetallic compound (IMC) on copper wire and aluminum pad bondability. *IEEE Trans. Compon. Packag. Technol.* **2003**, *26*, 367–374.
66. Uno, T. Bond reliability under humid environment for coated copper wire and bare copper wire. *Microelectron. Reliab.* **2010**, *51*, 148–156. [CrossRef]
67. Kenji, T.; Kazuya, F.; Syozo, M.; Takamichi, M. Development of Copper Wire Bonding Application Technology. *IEEE Trans. Comp. Hybrids Manufact. Technol.* **1990**, *13*, 667–672.
68. Murali, S.; Srikanth, N.; Vath, C.J. An analysis of intermetallics formation of gold and copper ball bonding on thermal aging. *Mater. Res. Bull.* **2003**, *38*, 637–646. [CrossRef]
69. Yuan, D.W.; Zeng, H.; Xiao, X.P.; Wang, H.; Han, B.J.; Liu, B.X.; Yang, B. Effect of Mg addition on Fe phase morphology, distribution and aging kinetics of Cu-6.5 Fe alloy. *Mater. Sci. Eng. A* **2021**, *812*, 141064. [CrossRef]
70. Goh, K.S.; Zhong, Z.W. A new bonding-tool solution to improve stitch bondability. *Microelectron. Eng.* **2007**, *84*, 173–179. [CrossRef]
71. Breach, C.D.; Wulff, F.W. A brief review of selected aspects of the materials science of ball bonding. *Microelectron. Reliab.* **2010**, *50*, 1–20. [CrossRef]
72. Shah, A.; Rockey, T.; Xu, H.; Qin, I.; Jie, W.; Yauw, O.; Chylak, B. Advanced wire bonding technology for Ag wire. In Proceedings of the 2015 17th Electronics Packaging Technology Conference (EPTC), Singapore, 2 December 2015.
73. Wu, X.Q.; Ye, D.H.; Zhang, H.M.; Song, L.; Guo, L.P. Improvement of inter layer dielectric crack for LQFP C90FG wafer technology devices in copper wire bonding process. *Microelectron. Int.* **2022**, *39*, 14–21. [CrossRef]
74. Li, J.H.; Liu, L.G.; Ma, B.K.; Deng, L.H.; Han, L. Dynamics Features of Cu-Wire Bonding during Overhang Bonding Process. *IEEE Electron. Device Lett.* **2011**, *32*, 1731–1733.
75. Chauhan, P.; Zhong, Z.W.; Pecht, M. Copper Wire Bonding Concerns and Best Practices. *J. Electron. Mater.* **2013**, *42*, 2415–2434. [CrossRef]
76. Uno, T. Enhancing bondability with coated copper bonding wire. *Microelectron. Reliab.* **2011**, *51*, 88–96. [CrossRef]
77. Zhang, G. Cu bonding wire technology for semiconductor packaging. *New Technol. New Prod. China* **2021**, *29*, 70–72.
78. Cao, J.; Fan, J.L.; Li, Y.M.; Wu, X.F. Corrosion-Resistant Bonding Copper Wire and Preparation Method Thereof. China Patent CN104278169A, 7 December 2013.
79. Fang, Y.B.; Zheng, Z.F.; Zheng, K.D. Bonded Copper Wire and Preparing Method Thereof. China Patent CN1949492A, 3 November 2006.
80. Yang, F.; Zhang, X.D.; Fang, F. Microstructure and properties of cold-drawn Cu and Cu-Fe alloy wires. *IOP Conf. Ser. Mater. Sci. Eng.* **2022**, *1249*, 012057. [CrossRef]
81. Lv, C.K.; Lv, C.C. Copper linking wire and production method thereof. China Patent CN101487090A, 17 January 2008.
82. Sato, M.; Suzuki, I.; Yamamori, K. Bonding Wire. Japan Patent JPH06168975A, 15 July 1993.
83. Sengul, S.; Guder, V. Key factors of deformation mechanism of Cu-Ag alloy. *J. Non-Cryst. Solids* **2022**, *576*, 121270. [CrossRef]
84. Zhu, X.F.; Xiao, Z.; An, J.H.; Jiang, H.Y.; Jiang, Y.B.; Li, Z. Microstructure and properties of Cu-Ag alloy prepared by continuously directional solidification. *J. Alloys Compd.* **2021**, *883*, 160769. [CrossRef]
85. Zhao, Z.L.; Xiao, Z.; Li, Z.; Qiu, W.T.; Jiang, H.Y.; Lei, Q.; Liu, Z.; Jiang, Y.B.; Zhang, S.J. Microstructure and properties of a Cu-Ni-Si-Co-Cr alloy with high strength and high conductivity. *Mater. Sci. Eng. A.* **2019**, *759*, 396–403. [CrossRef]
86. Zhu, X.F.; Xiao, Z.; Song, K.X.; Sheng, X.Y.; Dai, J.; Jiang, H.B.; Zhou, S.Q. Microstructure and properties evolution of Cu-Ag alloy prepared by downward continuous casting. *Chin. J. Nonferrous Met.* **2021**, *31*, 1176–1187.
87. Kong, L.B.; Zhou, Y.J.; Song, K.X.; Cao, J.; Lv, C.C.; Li, K.; Liu, Q.L.; Wu, B.A.; Tang, H.Y.; Zhang, X.B.; et al. Microstructure and Properties of Cu-Ag Alloy after Annealing at Different Temperatures. *Mater. Mech. Eng.* **2020**, *44*, 29–32.
88. Zhao, H.M.; Fu, H.D.; Xie, M.; Xie, J.X. Effect of Ag content and drawing strain on microstructure and properties of directionally solidified Cu-Ag alloy. *Vacuum* **2018**, *154*, 190–199. [CrossRef]
89. Xie, M.W.; Huang, W.; Chen, H.M.; Gong, L.K.; Xie, W.B.; Wang, H.; Yang, B. Microstructural evolution and strengthening mechanisms in cold-rolled Cu–Ag alloys. *J. Alloys Compd.* **2021**, *851*, 156893. [CrossRef]
90. Chen, M.H.; Wu, Y.; Zhang, W.H. Analysis on Quality Problem and Preventive Measure of Copper Rod Drawing. *Nonferrous Met. Process.* **2011**, *40*, 31–34.
91. Application of New Technologies in the Drawing of Coated Wire. Available online: https://www.wirecable.in/2016/02/application-of-new-technologies-in-the-drawing-of-coated-wire/ (accessed on 10 February 2016).

92. Du, Y.H.; Liu, Z.Q.; Ji, H.J.; Li, M.Y.; Wen, M. The mechanism of Pd distribution in the process of FAB formation during Pd-coated Cu wire bonding. *J. Mater. Sci. Mater. Electron.* **2018**, *29*, 13774–13781. [CrossRef]
93. Clauberg, H.; Chylak, B.; Wong, N.; Yeung, J.; Milke, E. Wire bonding with Pd-coated copper wire. In Proceedings of the IEEE 2010 IEEE CPMT Symposium Japan, Kyoto, Japan, 24 August 2010.
94. Qin, W.T.; Anderson, T.; Chang, G. Mechanism to improve the reliability of copper wire bonding with palladium-coating of the wire. *Microelectron. Reliab.* **2019**, *99*, 239–244. [CrossRef]
95. Singh, I.; Qin, I.; Xu, H.; Huynh, C.; Low, S.; Clauberg, H.; Chylak, B.; Acoff, V.L. Pd-coated Cu wire bonding technology: Chip design, process optimization, production qualification and reliability test for high reliability semiconductor devices. In Proceedings of the IEEE 2012 IEEE 62nd Electronic Components and Technology Conference (ECTC), San Diego, CA, USA, 29 May 2012.
96. Chen, K.J.; Hung, F.Y.; Chang, C.Y. A Study of the Sulfidation Behavior on Palladium-Coated Copper Wire with a Flash-Gold Layer (PCA) after Wire Bonding. *Electronics* **2019**, *8*, 792. [CrossRef]
97. Lau, K.T.; Cha, C.L. A Review of Palladium Coated Copper Wire Bonding for Automotive Device. *Int. J. Mech. Prod. Eng. Res. Dev.* **2020**, *10*, 4479–4492.
98. Zheng, K.D.; Feng, X.L.; Li, C.L. Palladium-plated bonded copper wire and production method thereof. China Patent CN101707194A, 11 November 2009.
99. Fang, Y.B. Surface Palladium-Plated Bonding Brass Wire. China Patent CN102130067B, 31 December 2010.
100. Fan, J.L.; Cao, J.; Gao, W.B.; Liu, Z.Q. Effect of coating speed and annealing on properties of direct Pd-coated copper bonding wire. *Heat Treat. Met.* **2016**, *41*, 89–93.
101. Liu, X.J.; Zhang, B.H.; Wang, D.J.; Cong, Y.Q.; Wang, J.Y. Properties of Cu-Al Intermetallic Compounds in Copper Wire Bonding. *Semicond. Technol.* **2011**, *36*, 880–884.
102. Mokhtari, O.; Nishikawa, H. Effect of surface potential distribution on corrosion behavior of Cu/Al interface in Cu wire bonding applications. *Microelectron. Reliab.* **2020**, *113*, 113942. [CrossRef]
103. Gan, C.L.; Hashim, U. Reliability Assessment and Activation Energy Study of Au and Pd-Coated Cu Wires Post High Temperature Aging in Nanoscale Semiconductor Packaging. *J. Electron. Packag.* **2013**, *135*, 210101–210107. [CrossRef]
104. Xu, H.; Hang, C.J.; Wang, C.Q.; Tian, Y.H. Contrast Research on the Reliability of Gold and Copper Wire/Ball Bonding. *Equip. Electron. Prod. Manuf.* **2006**, *5*, 23–27.
105. Xu, H.; Liu, C.; Silberschmidt, V.V.; Pramana, S.S.; White, T.J.; Chen, Z.; Acoff, V.L. Behavior of aluminum oxide, intermetallics and voids in Cu–Al wire bonds. *Acta Mater.* **2011**, *59*, 5661–5673. [CrossRef]
106. Jin, O.; Koizumi, M.; Araki, I. Investigation of the Reliability of Copper Ball bonds to Aluminum Electrodes. *IEEE Trans. Comp. Hybrids Manufact. Technol.* **1987**, *12*, 550–555.
107. Guo, Y.J.; Liu, G.W.; Jin, H.Y. Intermetallic phase formation in diffusion-bonded Cu/Al laminates. *J. Mater. Sci.* **2011**, *46*, 2467–2473. [CrossRef]
108. Xu, H.; Liu, C.; Silberschmidt, V.V.; Pramana, S.S.; White, T.J.; Chen, Z. A re-examination of the mechanism of thermosonic copper ball bonding on aluminium metallization pads. *Scr. Mater.* **2009**, *61*, 165–168. [CrossRef]
109. Chen, C.Y.; Hwang, W.S. Effect of Annealing on the Interfacial Structure of Aluminum-Copper Joints. *Mater. Trans.* **2007**, *48*, 1938–1947. [CrossRef]
110. Wu, S.P.; Cai, X.L.; Zhou, L.; Yang, C.J.; Li, W.H.; Cheng, Y.C. Contribution of Mechanical Activation on the Growth of Intermetallic Compound Layers at the Cu/Al Interface during Vacuum Hot Pressing. *Trans. Indian Inst. Met.* **2022**, *75*, 2129–2137. [CrossRef]
111. Lim, A.B.Y.; Boothroyd, C.B.; Yauw, O.; Chylak, B.; Gan, C.L.; Chen, Z. Interfacial evolution and bond reliability in thermosonic Pd coated Cu wire bonding on aluminum metallization: Effect of palladium distribution. *Microelectron. Reliab.* **2016**, *4*, 3. [CrossRef]
112. Chen, H.; Ricky Lee, S.W.; Ding, Y.T. Evaluation of Bondability and Reliability of Single Crystal Copper Wire Bonding. In Proceedings of the IEEE 2005 Conference on High Density Microsystem Design and Packaging and Component Failure Analysis, Shanghai, China, 27 June 2005.
113. Yuan, B.; Luo, Z.; Zhu, M.; Xu, Y.G.; Peng, S.Y. High-Reliability Copper Alloy Bonding Wire for Electronic Packaging and Preparation Method Therefor. U.S. Patent US20200373272A1, 26 November 2020.
114. Murali, S.; Yeung, J.; Perez, R. Alloyed Copper Bonding Wire with Homogeneous Microstructure. In Proceedings of the IEEE 2012 35th IEEE/CPMT International Electronics Manufacturing Technology Conference (IEMT), Ipoh, Malaysia, 6–8 November 2012.
115. Kaimori, S.; Nonaka, T.; Mizoguchi, A. The Development of Cu Bonding Wire with Oxidation-Resistant Metal Coating. *IEEE Trans. Adv. Packag.* **2006**, *29*, 227–231. [CrossRef]
116. Tajedini, M.; Osmanson, A.T.; Kim, Y.R.; Madanipour, H.; Kim, C.U.; Glasscock, B.; Khan, M. Electromigration Effect on the Pd Coated Cu Wirebond. In Proceedings of the IEEE 71st Electronic Components and Technology Conference (ECTC), San Diego, CA, USA, 1 June–4 July 2021.
117. Lim, A.B.Y.; Chang, A.C.K.; Lee, C.X.; Yauw, O.; Chylak, B.; Chen, Z. Palladium-coated and Bare Copper Wire Study for Ultra-Fine Pitch Wire Bonding. *ECS Trans. Adv.* **2013**, *52*, 717–730. [CrossRef]
118. Qin, I.; Xu, H.; Milton, B.; Clauberg, H.; Chylak, B. Molded Reliability Study for Different Cu Wire Bonding Configurations. In Proceedings of the IEEE 2013 IEEE 63th Electronic Components and Technology Conference (ECTC), Las Vegas, NV, USA, 28–31 May 2013.

119. Lim, A.; Neo, W.J.; Yauw, O.; Chylak, B.; Gan, C.L.; Chen, Z. Evaluation of the corrosion performance of Cu-Al intermetallic compounds and the effect of Pd addition. *Microelectron. Reliab.* **2016**, *56*, 155–161. [CrossRef]
120. Eto, M.; Araki, N.; Yamada, T.; Sugiyama, M.; Fujimoto, S. Influence of post-bonding heating process on the long-term reliability of Cu/Al contact. *Microelectron. Reliab.* **2021**, *118*, 114058. [CrossRef]
121. Jeon, J.; Na, S.H.; Jeon, S.H.; Mo, M.; Kang, D.B.; Lim, K.; Kim, J.Y. High Reliability Challenges with Cu Wire Bonding for Automotive Devices in the AEC-Q006. In Proceedings of the IEEE 2017 IEEE 67th Electronic Components and Technology Conference (ECTC), Orlando, FL, USA, 30 May–2 June 2017.
122. Eto, M.; Araki, N.; Yamada, T.; Klengel, R.; Klengel, S.; Petzold, M.; Sugiyama, M.; Fujimoto, S. Effects of alloying elements in high reliability copper wire bond material for high temperature applications. *Microelectron. Reliab.* **2020**, *114*, 113819. [CrossRef]
123. Sun, X.T.; Chen, W.J.; Xiong, X.Y.; Chen, W.H.; Jin, Y. A Variable Configuration Force Sensor with Adjustable Resolution for Robotic Applications. *IEEE Trans. Ind. Electron.* **2023**, *70*, 2066–2075. [CrossRef]
124. Hu, G.J. Comparison of copper, silver and gold wire bonding on interconnect metallization. In Proceedings of the IEEE 2012 13th International Conference on Electronic Packaging Technology & High Density Packaging (ICEPT-HDP), Guilin, China, 13–16 August 2012.
125. Yang, G.N.; Zhou, Z.Q.; Zhang, H.D.; Zhang, Y.; Peng, Z.; Gong, P.; Wang, X.; Cui, C.Q. Improved Anti-Vulcanization and Bonding Performance of a Silver Alloy Bonding Wire by a Cathodic Passivation Treatment with Palladium. *Materials* **2022**, *15*, 2355. [CrossRef]
126. Lu, C. Review on silver wire bonding. In Proceedings of the IEEE 2013 8th International Microsystems, Packaging, Assembly and Circuits Technology Conference (IMPACT), Taiwan, China, 22–25 December 2013.
127. Ren, Z. Silver bonding wire for BSOB(Bond-Stitch-on-Ball)/BBOS(Bonding-Ball-on-Stitch). In Proceedings of the IEEE 2016 China Semiconductor Technology International Conference (CSTIC), Shanghai, China, 13–14 March 2016.
128. Du, Y.H.; Liu, Z.Q.; Xiong, T.; Yuan, Z.B.; Leng, F.; Lv, H.L.; Yu, D.Q. Bond reliability under humid environment for Pd coated Cu and Ag alloy wire bonding. In Proceedings of the IEEE 2015 16th International Conference on Electronic Packaging Technology (ICEPT), Changsha, China, 11–14 August 2015.
129. Tanna, S.; Pisigan, J.L.; Song, W.H.; Halmo, C.; Persic, J.; Mayer, M. Low cost Pd coated Ag bonding wire for high quality FAB in air. In Proceedings of the IEEE 2012 IEEE 62nd Electronic Components and Technology Conference (ECTC), San Diego, CA, USA, 29 May–1 June 2012.
130. Xi, J.Q.; Mendoza, N.; Chen, K.; Yang, T.; Reyes, E.; Bezuk, S.; Lin, J.; Ke, S.G.; Chen, E. Evaluation of Ag Wire Reliability on Fine Pitch Wire Bonding. In Proceedings of the IEEE 2015 IEEE 65th Electronic Components and Technology Conference (ECTC), San Diego, CA, USA, 26–29 May 2015.
131. Chuang, T.H.; Lee, P.I.; Lin, Y.C. An Optimized Ag–5Pd–3.5Au Bonding Wire for the Resistance of Ag Ion Migration in LED Packages. *IEEE Trans. Compon. Packag. Manuf. Technol.* **2020**, *10*, 1989–1995. [CrossRef]
132. Tseng, Y.W.; Hung, F.Y.; Lui, T.S. Microstructure, tensile and electrical properties of gold-coated silver bonding wire. *Microelectron. Reliab.* **2015**, *55*, 608–612. [CrossRef]
133. Wu, Y.H.; Hung, F.Y.; Lui, T.S.; Chen, K.J. Study of wire bonding reliability of Ag-Pd-Au alloy wire with flash-gold after chlorination and sulfidation. *Microelectron. Reliab.* **2019**, *99*, 183–196. [CrossRef]
134. Tseng, Y.W.; Hung, F.Y.; Lui, T.S. Microstructure, Mechanical and High-Temperature Electrical Properties of Cyanide-Free Au-Coated Ag Wire (ACA). *Mater. Trans.* **2015**, *56*, 441–444. [CrossRef]
135. Kang, F.F.; Zhou, W.Y.; Wu, Y.J. Microstructure and Properties of Gold-Coated Silver Composite Bonding Wire for Microelectronic Package. *Semicond. Technol.* **2018**, *43*, 702–707.
136. Liang, Z.; Xie, Y.H.; Yan, R.; Xiang, C.H.; Zhou, X.G.; Su, H.F. A Bonding Wire. China Patent CN105428335A, 23 March 2016.
137. Gao, M.Y.; Xie, H.B.; Fang, Y.T.; Wang, H.T.; Liu, J.B. Progress in surface treatment techniques of copper and copper alloys. *Chin. J. Nonferrous Met.* **2021**, *31*, 1121–1133.
138. Song, K.X.; Ding, Y.T.; Wu, B.A.; Hu, R.; Chen, D.B.; Lu, W.W.; Hu, H.; Zhou, Y.J. Controlling the wettability of Pd coating solution and its effect on the uniformity of coating layer. *J. Funct. Mater.* **2021**, *52*, 7060–7063.
139. Fan, J.L.; Zhu, L.X.; Cao, J.; Hua, H. Effect of Au component in Ag alloy bonding wire on bonded strength and bonded reliability. *Funct. Mater.* **2019**, *50*, 10145–10148.
140. Kuo, B.H.; Tsai, D.C.; Huang, Y.L.; Hsu, P.C.; Chuang, T.H.; Lee, J.D.; Tsai, H.H.; Shieu, F.S. Effect of alloying Au on the microstructural, mechanical and electrical properties of Ag-based alloy wires. *J. Mater. Sci. Mater. Electron.* **2019**, *30*, 9396–9409. [CrossRef]
141. Rosalbino, F.; Delsante, S.; Borzone, G.; Scavino, G. Influence of noble metals alloying additions on the corrosion behaviour of tita-nium in a fluoride-containing environment. *J. Mater. Sci. Mater. Med.* **2012**, *23*, 1129–1137. [CrossRef]
142. Tsai, H.H.; Lee, J.D.; Tsai, C.H.; Wang, H.C.; Chang, C.C.; Chuang, T.H. An innovative annealing-twinned Ag-Au-Pd bonding wire for IC and LED packaging. In Proceedings of the IEEE 2012 7th International Microsystems, Packaging, Assembly and Circuits Technology Conference (IMPACT), Taipei, China, 24–26 October 2012.
143. Tsai, C.H.; Chuang, C.H.; Tsai, H.H.; Lee, J.D.; Chang, D.; Lin, H.J.; Chuang, T.H. Materials Characteristics of Ag-Alloy Wires and Their Applications in Advanced Packages. *IEEE Trans. Compon. Packag. Manuf. Technol.* **2016**, *6*, 298–305. [CrossRef]
144. Cao, J.; Wu, W.X.; Zhang, C.; Hua, H.; Xu, X. Effects of heat treatment temperature on properties and microstructure of Ag-4Pd alloy bonding wire. *Trans. Mater. Heat Treat.* **2018**, *39*, 44–48.

145. Cao, J.; Fan, J.L.; Gao, W.B. Effects of Drawing and Annealing on Properties of Ag-4Pd Alloy Bonding Wire. *Chin. J. Mech.* **2016**, *52*, 92–97. [CrossRef]
146. Chuang, T.H.; Wang, H.C.; Chuang, C.H.; Lee, J.D.; Tsai, H.H. Effect of Annealing Twins on Electromigration in Ag-8Au-3Pd Bonding Wires. *J. Electron. Mater.* **2013**, *42*, 545–551. [CrossRef]
147. Chuang, T.H.; Tsai, C.H.; Wang, H.C.; Chang, C.C.; Chuang, C.H.; Lee, J.D.; Tsai, H.H. Effects of Annealing Twins on the Grain Growth and Mechanical Properties of Ag-8Au-3Pd Bonding Wires. *J. Electron. Mater.* **2012**, *41*, 3215–3222. [CrossRef]
148. Chuang, T.H.; Wang, H.C.; Tsai, C.H.; Chang, C.C.; Chuang, C.H.; Lee, J.D.; Tsai, H.H. Thermal stability of grain structure and material properties in an annealing-twinned Ag–8Au–3Pd alloy wire. *Scr. Mater.* **2012**, *67*, 605–608. [CrossRef]
149. Luechinger, C.; Chen, R.; Fu, J.; Poncelet, B.; Valentin, O.; Walker, T.J.; Xu, T. Aluminum–copper ribbon interconnects for power devices. *IEEE Trans. Compon. Packag. Manuf. Technol.* **2017**, *7*, 1567–1577. [CrossRef]
150. Chen, C.H.; Lin, Y.C.; Groth, A.; Lai, Y.C.; Lin, C.Y.; Chang, H.M.; Chuang, T.H. Ultrasonic Bonding of Ag and Ag-Alloy Ribbon—An Innovative Alternative for High Power IC Packages. *IEEE Trans. Compon. Packag. Manuf. Technol.* **2020**, *10*, 1061–1068. [CrossRef]
151. Chen, C.H.; Lin, Y.C.; Chuang, T.H. Grain growth and twin formation in a Ag-4Pd alloy ribbon after annealing treatments. *J. Alloys Compd.* **2021**, *863*, 158619. [CrossRef]
152. Chen, C.H.; Chuang, T.H. Optimization of Ag-alloy ribbon bonding—An approach to reliable interconnection for high power IC packaging. *Microelectron. Reliab.* **2022**, *137*, 114786. [CrossRef]
153. Hung, F.Y.; Lui, T.S.; Chen, L.H.; Lan, K.A. Microstructures and Fusing Electrical Current of Microelectronic Sn-9Zn-(0.25RE) Solders. *Mater. Trans.* **2008**, *49*, 1491–1495. [CrossRef]
154. Hsueh, H.W.; Hung, F.Y.; Lui, T.S.; Chen, L.H.; Chang, J.K. Microstructure, electric flame-off (EFO) characteristics and tensile properties of silver–lanthanum alloy wire. *Microelectron. Reliab.* **2014**, *54*, 2564–2569. [CrossRef]
155. Cao, J.; Fan, J.L.; Liu, Z.Q.; Zhang, Y.M. Effect of silver alloy bonding wire properties on bond strengths and reliability. In Proceedings of the ICEP-IAAC, Kyoto, Japan, 14 April 2015.
156. Cao, J.; Wu, W.X. Effects of Au coated Ag bonding wire properties on bonded quality. *Mater. Sci. Technol.* **2018**, *26*, 30–35.
157. Cao, J.; Zhang, J.C.; Persic, J.; Song, K.X. Effects of Bonding Parameters on Free Air Ball Properties and Bonded Strength of Ag-10Au-3.6Pd Alloy Bonding Wire. *Micromachines* **2020**, *11*, 777. [CrossRef]
158. Zhou, W.Y.; Wu, Y.J.; Chen, J.L.; Yang, G.X.; Kong, J.W.; Kang, F.F. Effect of Electronic Flame off Parameters on the Bonding Using Silve Wire. *Precious Met.* **2017**, *38*, 34–39.
159. Krumbein, S.J. Metallic Electromigration Phenomena. *IEEE Trans. Compon. Hybrids Manuf. Technol.* **1988**, *11*, 5–15. [CrossRef]
160. Chuang, C.J.; Lin, Y.C.; He, Y.Z.; Tsai, C.H.; Lee, J.D.; Wang, A.C.; Tsai, H.H. Inhibition of Silver Electrolytic Migration in Ag-alloy Bonding Wires. *Mater. Sci. Forum* **2016**, *863*, 95–101. [CrossRef]
161. Lin, Y.C.; Lee, P.L.; Wu, P.C.; Chen, C.H.; Chuang, T.H. Effects of Grain Size on the Ag Dissolution and Ion Migration of Ag-4Pd Alloy Wires. *J. Electron. Mater.* **2021**, *50*, 5955–5964. [CrossRef]
162. Xiao, Y.C.; Tang, H.Y.; Zhang, H.H.; Sun, L.; Xie, L.; Wu, B.A.; Xie, W.D.; Cai, X.N.; Li, F. Effect of Au Coating on the Electrolytic Migration of Ag Bonding Wires for Electronic Packaging Applications. *J. Phys. Conf. Ser.* **2022**, *2393*, 012013. [CrossRef]
163. Cho, J.S.; YOO, K.A.; Moon, J.T.; Son, S.B.; Lee, S.H.; Oh, K.H. Pd effect on reliability of Ag bonding wires in microelectronic devices in high-humidity environments. *Met. Mater. Int.* **2012**, *18*, 881–1885. [CrossRef]
164. Cai, X.H.; Li, J.D.; Lin, X.R. Study on Electromigration Failure of High Performance Ag-4Pd Alloy Wire. In Proceedings of the 2014 Cross-Strait Destruction Science and Materials Testing Academic Conference and the 12th Destruction Science Symposium/The 10th National MTS Materials Testing Academic Conference, Taiwan, China, 22 October 2014.
165. Guo, R.; Gao, L.M.; Li, M.; Mao, D.L.; Qian, K.Y.; Chiu, H. Microstructure evolution of Ag-8Au-3Pd alloy wire during electromigration. *Mater. Charact.* **2015**, *110*, 44–51. [CrossRef]
166. Chen, C.H.; Lee, P.I.; Chuang, T.H. Microstructure evolution and failure mechanism of electromigration in Ag-alloy bonding wire. *J. Alloys Compd.* **2022**, *913*, 165266. [CrossRef]
167. Peng, C.; Liang, S.; Huang, F.X.; Zhong, M.J.; Ran, X.J. Research Progress on Bonding Interface of Bonded Wire. *Mater. Rep.* **2019**, *33*, 501–504.
168. Hsu, T.Y.; Chang, J.Y.; Leu, F.J.; Chang, H.M.; Ouyang, F.Y. Ag alloy wire bonding under electromigration test. In Proceedings of the ICEP-IAAC, Kyoto, Japan, 14 April 2015.
169. Fu, S.W.; Lee, C.C. Solid-state reactions of silver and aluminum associated with silver wire bonds. In Proceedings of the 2016 IEEE 66th Electronic Components and Technology Conference, Las Vegas, NV, USA, 31 May–3 June 2016.
170. Ho, C.C.; Chen, K.J.; Hung, F.Y.; Lui, T.S. Study of electrical fatigue test in gold-coated siliver-4 wt.% palladium bonding wire. *Microelectron. Reliab.* **2019**, *103*, 113502. [CrossRef]
171. Chuang, T.H.; Chang, C.C.; Chuang, C.H.; Lee, J.D.; Tsai, H.H. Formation and Growth of Intermetallics in an Annealing-Twinned Ag-8Au-3Pd Wire Bonding Package during Reliability Tests. *IEEE Trans. Compon. Packag. Manuf. Technol.* **2013**, *3*, 3–9. [CrossRef]
172. Long, Y.Y.; Arndt, M.; Dencker, F.; Wurz, M.; Twiefel, J.; Wallaschek, J. Impact of surface texture on ultrasonic wire bonding process. *J. Mater. Res. Technol.* **2022**, *20*, 1828–1838. [CrossRef]
173. Zou, Y.S.; Gan, C.L.; Chung, M.; Takiar, H. A review of interconnect materials used in emerging memory device packaging: First- and second-level interconnect materials. *J. Mater. Sci. Mater. Electron.* **2021**, *32*, 27133–27147. [CrossRef]

174. Liang, S.; Huang, F.X.; Peng, C.; Zhong, M.J.; Wu, B.A.; Tang, H.Y. Research progress on copper and siliver bonding wires for microelectronic packaging technology. *J. Funct. Mater.* **2019**, *50*, 5048–5053.
175. Chen, X. Comparative study on bonding of palladium-coated copper wire and bare copper wire in IC package. *Silicon Val.* **2014**, *7*, 152–153.
176. Yu, C.F.; Chan, C.M.; Chan, L.C.; Hsieh, K.C. Cu wire bond microstructure analysis and failure mechanism. *Microelectron. Reliab.* **2011**, *51*, 119–124. [CrossRef]

MDPI
St. Alban-Anlage 66
4052 Basel
Switzerland
Tel. +41 61 683 77 34
Fax +41 61 302 89 18
www.mdpi.com

Micromachines Editorial Office
E-mail: micromachines@mdpi.com
www.mdpi.com/journal/micromachines

www.ingramcontent.com/pod-product-compliance
Lightning Source LLC
Chambersburg PA
CBHW062033220926
51850CB00010B/2530
* 9 7 8 3 0 3 6 5 8 2 1 3 9 *